22.

Holt Mathematics

AF860589

Course 3

Homework and Practice Workbook

Teacher's Guide

WASHOE COUNTY SCHOOL DISTRICT
GIFTED & TALENTED PROGRAM

HOLT, RINEHART AND WINSTON
A Harcourt Education Company
Orlando • Austin • New York • San Diego • London

22.

Copyright © by Holt, Rinehart and Winston

All rights reserved. No part of this publication may be reproduced or transmitted in any form or by any means, electronic or mechanical, including photocopy, recording, or any information storage and retrieval system, without permission in writing from the publisher.

Teachers using HOLT MATHEMATICS may photocopy complete pages in sufficient quantities for classroom use only and not for resale.

Printed in the United States of America

If you have received these materials as examination copies free of charge, Holt, Rinehart and Winston retains title to the materials and they may not be resold. Resale of examination copies is strictly prohibited and is illegal.

Possession of this publication in print format does not entitle users to convert this publication, or any portion of it, into electronic format.

ISBN 0-03-078466-2

4 5 018 09 08

LESSON 1-1

Practice

Variables and Expressions

Evaluate each expression for the given value of the variable.

1. $6x + 2$ for $x = 3$ — 20
2. $18 - a$ for $a = 13$ — 5
3. $\frac{1}{4}y$ for $y = 16$ — 4
4. $9 - 2b$ for $b = 3$ — 3
5. $44 - 12n$ for $n = 3$ — 8
6. $7.2 + 8k$ for $k = 2$ — 23.2
7. $20(b - 15)$ for $b = 19$ — 80
8. $n(18 - 5)$ for $n = 4$ — 52

Evaluate each expression for the given value of the variables.

9. $2x + y$ for $x = 7$ and $y = 11$ — 25
10. $4j - k$ for $j = 4$ and $k = 10$ — 6
11. $9a - 6b$ for $a = 6$ and $b = 2$ — 42
12. $5s + 5t$ for $s = 15$ and $t = 12$ — 135
13. $7(n - m)$ for $m = 4$ and $n = 15$ — 77
14. $w(14 - y)$ for $w = 8$ and $y = 5$ — 72

If q is the number of quarts of lemonade, then $\frac{1}{4}q$ can be used to find the number of cups of lemonade mix needed to make the lemonade. How much mix is needed to make each amount of lemonade?

15. 2 quarts — $\frac{1}{2}$ cup
16. 8 quarts — 2 cups
17. 12 quarts — 3 cups
18. 18 quarts — $4\frac{1}{2}$ cups

19. If m is the number of minutes a taxi ride lasts, then $2 + 0.35m$ can be used to find the cost of a taxi ride with Bill's Taxi Company.

How much will it cost for a 12-min taxi ride? $6.20

Copyright © by Holt, Rinehart and Winston. All rights reserved.

LECCIÓN 1-1

Práctica

Variables y expresiones

Evalúa cada expresión para el valor dado de la variable.

1. $6x + 2$ para $x = 3$ — 20
2. $18 - a$ para $a = 13$ — 5
3. $\frac{1}{4}y$ para $y = 16$ — 4
4. $9 - 2b$ para $b = 3$ — 3
5. $44 - 12n$ para $n = 3$ — 8
6. $7.2 + 8k$ para $k = 2$ — 23.2
7. $20(b - 15)$ para $b = 19$ — 80
8. $n(18 - 5)$ para $n = 4$ — 52

Evalúa cada expresión para el valor dado de la variable.

9. $2x + y$ para $x = 7$ y $y = 11$ — 25
10. $4j - k$ para $j = 4$ y $k = 10$ — 6
11. $9a - 6b$ para $a = 6$ y $b = 2$ — 42
12. $5s + 5t$ para $s = 15$ y $t = 12$ — 135
13. $7(n - m)$ para $m = 4$ y $n = 15$ — 77
14. $w(14 - y)$ para $w = 8$ y $y = 5$ — 72

Si c es la cantidad de cuartos de limonada, entonces se puede usar $\frac{1}{4}c$ para hallar la cantidad de tazas de mezcla de limonada que se necesitan para prepararla. ¿Cuánta mezcla se necesita para preparar cada cantidad de limonada?

15. 2 cuartos — $\frac{1}{2}$ taza
16. 8 cuartos — 2 tazas
17. 12 cuartos — 3 tazas
18. 18 cuartos — $4\frac{1}{2}$ tazas

19. Si m es la cantidad de minutos que dura un viaje en taxi, entonces se puede usar $2 + 0.35m$ para hallar el costo de un viaje en un taxi de la compañía de taxis de Bill.

¿Cuánto costará un viaje de 12 minutos? $6.20

Copyright © by Holt, Rinehart and Winston. All rights reserved.

LESSON 1-2

Practice

Algebraic Expressions

Write an algebraic expression for each word phrase.

1. 6 less than twice x — $2x - 6$
2. 1 more than the quotient of 21 and b — $1 + \frac{21}{b}$
3. 3 times the sum of b and 5 — $3(b + 5)$
4. 10 times the difference of d and 3 — $10(d - 3)$
5. the sum of 11 times s and 3 — $11s + 3$
6. 7 minus the product of 2 and x — $7 - 2x$

Write a word phrase for each algebraic expression.
Phrases may vary. Possible answers are shown.

7. $2n + 4$ — 4 more than twice n
8. $3r - 1$ — 1 less than the product of 3 and r
9. $10 - 6n$ — the difference of 10 and 6 times n
10. $7 + \frac{2}{c}$ — 7 more than 2 divided by c
11. $15x - 12$ — 12 less than the product of 15 and x
12. $\frac{y}{5} + 8$ — 8 more than the quotient of y and 5

13. Maddie earns $8 per hour. Write an algebraic expression to evaluate how much money Maddie will earn if she works for 15, 20, 25, or 30 hours.

n	$8n$	Earnings
15	8(15)	$120
20	8(20)	$160
25	8(25)	$200
30	8(30)	$240

14. Write a word problem that can be evaluated by the algebraic expression $y - 95$, and evaluate it for $y = 125$.

Possible answer: Marco has saved y dollars.
He wants to buy a skateboard that costs $95.
How much money will Marco have left after he buys the skateboard? $30

Copyright © by Holt, Rinehart and Winston. All rights reserved.

LECCIÓN 1-2

Práctica

Expresiones algebraicas

Escribe una expresión algebraica para cada expresión con palabras.

1. 6 menos que el doble de x — $2x - 6$
2. 1 más que el cociente de 21 y b — $1 + \frac{21}{b}$
3. 3 por la suma de b y 5 — $3(b + 5)$
4. 10 por la diferencia de d y 3 — $10(d - 3)$
5. la suma de 11 por s y 3 — $11s + 3$
6. 7 menos el producto de 2 y x — $7 - 2x$

Escribe una expresión con palabras para cada expresión algebraica.
Las expresiones variarán. Se muestran respuestas posibles.

7. $2n + 4$ — 4 más que el doble de n
8. $3r - 1$ — 1 menos que el producto de 3 y r
9. $10 - 6n$ — la diferencia de 10 y 6 por n
10. $7 + \frac{2}{c}$ — 7 más que 2 dividido entre c
11. $15x - 12$ — 12 menos que el producto de 15 y x
12. $\frac{y}{5} + 8$ — 8 más que el cociente de y y 5

13. Maddie gana $8 por hora. Escribe una expresión algebraica para evaluar cuánto dinero ganará Maddie si trabaja 15, 20, 25 ó 30 horas.

n	$8n$	Ganancias
15	8(15)	$120
20	8(20)	$160
25	8(25)	$200
30	8(30)	$240

14. Escribe un problema con palabras que se pueda evaluar mediante la expresión algebraica $y - 95$ y evalúalo para $y = 125$.

Respuesta posible: Marco ha ahorrado y dólares.
Quiere comprar una patineta que cuesta $95.
¿Cuánto dinero le quedará a Marco después de comprar la patineta? $30

Copyright © by Holt, Rinehart and Winston. All rights reserved.

Copyright © by Holt, Rinehart and Winston. All rights reserved.

LESSON **1-3**
Practice
Integers and Absolute Value

Write the integers in order from least to greatest.

1. 7, 3, −9 — −9, 3, 7
2. −6, 2, −5 — −6, −5, 2
3. −4, 1, −1 — −4, −1, 1
4. −8, 2, −11 — −11, −8, 2
5. −12, −15, 0 — −15, −12, 0
6. −24, −17, 30 — −24, −17, 30
7. 16, −14, −7 — −14, −7, 16
8. −9, −7, −16 — −16, −9, −7
9. −19, −23, −10 — −23, −19, −10

Find the additive inverse of each integer.

10. −8 — 8
11. 6 — −6
12. −14 — 14
13. 29 — −29

Evaluate each expression.

14. $|-8| + |-4|$ — 12
15. $|-12| + |12|$ — 24
16. $|19| + |-8|$ — 27
17. $|29 - 16|$ — 13
18. $|35 - 9|$ — 26
19. $|14 - 14|$ — 0
20. $|-15| + |10|$ — 25
21. $|-9| + |30|$ — 39
22. $|24| + |-8|$ — 32

23. Natalie keeps track of her bowling scores. The scores for the games she played this Saturday relative to her best score last Saturday are Game A, 6; Game B, -3; Game C, 8; and Game D, −5. Use <, >, or = to compare her first two games. Then list her games in order from the lowest score to the highest.

$6 > -3$; Game D, Game B, Game A, Game C

Copyright © by Holt, Rinehart and Winston. All rights reserved.

LECCIÓN **1-3**
Práctica
Enteros y valor absoluto

Escribe los enteros en orden de menor a mayor.

1. 7, 3, −9 — −9, 3, 7
2. −6, 2, −5 — −6, −5, 2
3. −4, 1, −1 — −4, −1, 1
4. −8, 2, −11 — −11, −8, 2
5. −12, −15, 0 — −15, −12, 0
6. −24, −17, 30 — −24, −17, 30
7. 16, −14, −7 — −14, −7, 16
8. −9, −7, −16 — −16, −9, −7
9. −19, −23, −10 — −23, −19, −10

Halla el inverso aditivo de cada entero.

10. −8 — 8
11. 6 — −6
12. −14 — 14
13. 29 — −29

Evalúa cada expresión.

14. $|-8| + |-4|$ — 12
15. $|-12| + |12|$ — 24
16. $|19| + |-8|$ — 27
17. $|29 - 16|$ — 13
18. $|35 - 9|$ — 26
19. $|14 - 14|$ — 0
20. $|-15| + |10|$ — 25
21. $|-9| + |30|$ — 39
22. $|24| + |-8|$ — 32

23. Natalie lleva un registro de su puntaje en el boliche. Los puntajes que obtuvo en los partidos de este sábado en relación con su mejor puntaje del sábado pasado son: partido A, 6; partido B, −3; partido C, 8; y partido D, −5. Usa < >, ó = para comparar los primeros dos partidos. Luego, haz una lista de los partidos del puntaje menor al mayor.

$6 > -3$; Partido D, Partido B, Partido A, Partido C

Copyright © by Holt, Rinehart and Winston. All rights reserved.

LESSON **1-4**
Practice
Adding Integers

Use a number line to find each sum.

1. 3 + 1 — 4

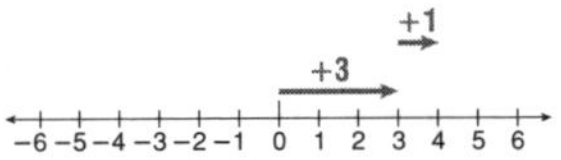

2. −3 + 2 — −1

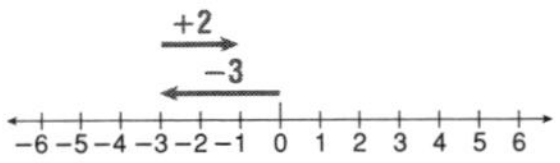

Add.

3. −5 + 18 — 13
4. −10 + 17 — 7
5. −22 + (−9) — −31
6. 24 + (−15) — 9

Evaluate each expression for the given value of the variable.

7. $r + 7$ for $r = 3$ — 10
8. $m + 5$ for $m = 9$ — 14
9. $x + 9$ for $x = 4$ — 13
10. $-6 + t$ for $t = -8$ — −14
11. $-7 + y$ for $y = -4$ — −11
12. $x + 9$ for $x = -8$ — 1
13. $-5 + d$ for $d = -2$ — −7
14. $x + (-4)$ for $x = -4$ — −8
15. $k + (-3)$ for $k = -5$ — −8
16. $-8 + b$ for $b = 13$ — 5
17. $-10 + d$ for $d = -2$ — −12
18. $t + (-3)$ for $t = 3$ — 0

19. Joleen has 2560 trading cards in her collection. She buys 165 new cards for the collection. How many trading cards does she have now?

2725 trading cards

20. The running back for the Bears carries the ball twice in the first quarter. The first run he gained fifteen yards and the second run he lost eight yards. How many yards did the two runs total?

7 yards

Copyright © by Holt, Rinehart and Winston. All rights reserved.

LECCIÓN **1-4**
Práctica
Cómo sumar enteros

Usa una recta numérica para hallar cada suma.

1. 3 + 1 — 4

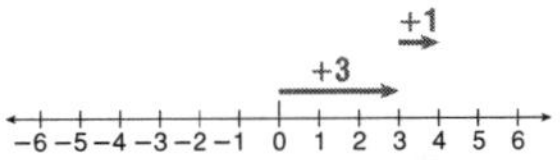

2. −3 + 2 — −1

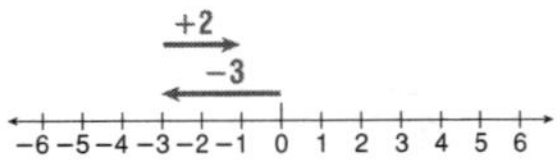

Suma.

3. −5 + 18 — 13
4. −10 + 17 — 7
5. −22 + (−9) — −31
6. 24 + (−15) — 9

Evalúa cada expresión para el valor dado de la variable.

7. $r + 7$ para $r = 3$ — 10
8. $m + 5$ para $m = 9$ — 14
9. $x + 9$ para $x = 4$ — 13
10. $-6 + t$ para $t = -8$ — −14
11. $-7 + y$ para $y = -4$ — −11
12. $x + 9$ para $x = -8$ — 1
13. $-5 + d$ para $d = -2$ — −7
14. $x + (-4)$ para $x = -4$ — −8
15. $k + (-3)$ para $k = -5$ — −8
16. $-8 + b$ para $b = 13$ — 5
17. $-10 + d$ para $d = -2$ — −12
18. $t + (-3)$ para $t = 3$ — 0

19. Joleen tiene una colección de 2560 estampas. Compra 165 estampas nuevas para la colección. ¿Cuántas tiene ahora?

2725 estampas

20. El corredor de los Bears lleva la pelota dos veces en el primer cuarto. En la primera corrida gana quince yardas y en la segunda pierde 8 yardas. ¿Cuánto suman las dos corridas?

7 yardas

Copyright © by Holt, Rinehart and Winston. All rights reserved.

Copyright © by Holt, Rinehart and Winston. All rights reserved.

LESSON **1-5**
Practice
Subtracting Integers

Subtract.

1. $8 - 2$ — 6
2. $10 - 5$ — 5
3. $7 - 12$ — -5
4. $16 - 10$ — 6
5. $3 - 10$ — -7
6. $16 - 9$ — 7
7. $-4 - 9$ — -13
8. $-8 - 10$ — -18
9. $33 - 57$ — -24
10. $16 - 49$ — -33
11. $-114 - 19$ — -133
12. $-88 - (-10)$ — -78

Evaluate each expression for the given value of the variable.

13. $x - 8$ for $x = 10$ — 2
14. $w - 10$ for $w = 15$ — 5
15. $15 - w$ for $w = 8$ — 7
16. $12 - t$ for $t = -8$ — 20
17. $15 - x$ for $x = -12$ — 27
18. $w - 20$ for $w = -15$ — -35
19. $-15 - x$ for $x = -10$ — -5
20. $-9 - x$ for $x = -20$ — 11
21. $-11 - d$ for $d = -15$ — 4
22. $y - (-10)$ for $y = -10$ — 0
23. $x - (-15)$ for $x = -5$ — 10
24. $a - (-12)$ for $a = 10$ — 22

25. The altitude of Mt. Blackburn in Alaska is 16,390 feet. The altitude of Mt. Elbert in Colorado is 14,433 feet. What is the difference in the altitudes of the two mountains?

1957 feet

26. In January, Jesse weighed 230 pounds. By November, he weighed 185 pounds. How much did Jesse's weight change?

-45 pounds

Copyright © by Holt, Rinehart and Winston. All rights reserved.

LECCIÓN **1-5**
Práctica
Cómo restar enteros

Resta.

1. $8 - 2$ — 6
2. $10 - 5$ — 5
3. $7 - 12$ — -5
4. $16 - 10$ — 6
5. $3 - 10$ — -7
6. $16 - 9$ — 7
7. $-4 - 9$ — -13
8. $-8 - 10$ — -18
9. $33 - 57$ — -24
10. $16 - 49$ — -33
11. $-114 - 19$ — -133
12. $-88 - (-10)$ — -78

Evalúa cada expresión para el valor dado de la variable.

13. $x - 8$ para $x = 10$ — 2
14. $w - 10$ para $w = 15$ — 5
15. $15 - w$ para $w = 8$ — 7
16. $12 - t$ para $t = -8$ — 20
17. $15 - x$ para $x = -12$ — 27
18. $w - 20$ para $w = -15$ — -35
19. $-15 - x$ para $x = -10$ — -5
20. $-9 - x$ para $x = -20$ — 11
21. $-11 - d$ para $d = -15$ — 4
22. $y - (-10)$ para $y = -10$ — 0
23. $x - (-15)$ para $x = -5$ — 10
24. $a - (-12)$ para $a = 10$ — 22

25. El monte Blackburn, ubicado en Alaska, mide 16,390 pies. El monte Elbert, ubicado en Colorado, mide 14,433 pies. ¿Qué diferencia de altura hay entre las dos montañas?

1957 pies

26. En enero, Jessie pesaba 230 libras. En noviembre, pesaba 185 libras. ¿Cuánto cambió el peso de Jessie?

-45 libras

Copyright © by Holt, Rinehart and Winston. All rights reserved.

LESSON **1-6**
Practice
Multiplying and Dividing Integers

Multiply or divide.

1. $6 \cdot 7$ — 42
2. $\frac{-15}{5}$ — -3
3. $-7 \cdot 3$ — -21
4. $\frac{20}{-4}$ — -5
5. $\frac{-36}{-4}$ — 9
6. $-8(-9)$ — 72
7. $\frac{-48}{-6}$ — 8
8. $7(-7)$ — -49
9. $5(-8)$ — -40
10. $(-6)(-9)$ — 54
11. $\frac{-36}{4}$ — -9
12. $\frac{42}{-7}$ — -6
13. $-9(-3)$ — 27
14. $(-4)(8)$ — -32
15. $\frac{-54}{-9}$ — 6
16. $\frac{-72}{8}$ — -9

Simplify.

17. $-5(3 + 7)$ — -50
18. $10(8 - 2)$ — 60
19. $-4(12 - 3)$ — -36
20. $9(15 - 8)$ — 63
21. $12(-9 + 4)$ — -60
22. $-11(7 - 13)$ — 66
23. $15(-12 + 8)$ — -60
24. $-10(-8 - 6)$ — 140
25. $6(-12 + 1)$ — -66
26. $-5(3 - 12)$ — 45
27. $-8(-5 - 5)$ — 80
28. $7(12 - 3)$ — 63
29. $10(-7 - 1)$ — -80
30. $12(2 - 5)$ — -36
31. $-15(-2 - 1)$ — 45
32. $9(8 - 20)$ — -108

33. Kristin and her three friends buy a pizza with twelve slices and split it equally. How many slices will each person receive?

3 slices

34. The temperature was -1°F, -5°F, 8°F, and -6°F on four consecutive days. What was the average temperature for those days?

-1°F

Copyright © by Holt, Rinehart and Winston. All rights reserved.

LECCIÓN **1-6**
Práctica
Cómo multiplicar y dividir enteros

Multiplica o divide.

1. $6 \cdot 7$ — 42
2. $\frac{-15}{5}$ — -3
3. $-7 \cdot 3$ — -21
4. $\frac{20}{-4}$ — -5
5. $\frac{-36}{-4}$ — 9
6. $-8(-9)$ — 72
7. $\frac{-48}{-6}$ — 8
8. $7(-7)$ — -49
9. $5(-8)$ — -40
10. $(-6)(-9)$ — 54
11. $\frac{-36}{4}$ — -9
12. $\frac{42}{-7}$ — -6
13. $-9(-3)$ — 27
14. $(-4)(8)$ — -32
15. $\frac{-54}{-9}$ — 6
16. $\frac{-72}{8}$ — -9

Simplifica.

17. $-5(3 + 7)$ — -50
18. $10(8 - 2)$ — 60
19. $-4(12 - 3)$ — -36
20. $9(15 - 8)$ — 63
21. $12(-9 + 4)$ — -60
22. $-11(7 - 13)$ — 66
23. $15(-12 + 8)$ — -60
24. $-10(-8 - 6)$ — 140
25. $6(-12 + 1)$ — -66
26. $-5(3 - 12)$ — 45
27. $-8(-5 - 5)$ — 80
28. $7(12 - 3)$ — 63
29. $10(-7 - 1)$ — -80
30. $12(2 - 5)$ — -36
31. $-15(-2 - 1)$ — 45
32. $9(8 - 20)$ — -108

33. Kristin y tres amigas compran una pizza de doce porciones y la dividen en partes iguales. ¿Cuántas porciones recibirá cada una?

3 porciones

34. La temperatura durante cuatro días consecutivos fue de -1° F, -5° F, 8° F y -6° F. ¿Cuál fue el promedio de temperatura de esos cuatro días?

-1°F

Copyright © by Holt, Rinehart and Winston. All rights reserved.

Copyright © by Holt, Rinehart and Winston.
All rights reserved.

LESSON 1-7

Practice

Solving Equations by Adding or Subtracting

Determine which value is a solution of the equation.

1. $x - 6 = 12$; $x = 6, 8,$ or 18 — $x = 18$
2. $9 + x = 17$; $x = 6, 8,$ or 26 — $x = 8$
3. $x - 12 = 26$; $x = 14, 38,$ or 40 — $x = 38$
4. $x + 18 = 59$; $x = 37, 41,$ or 77 — $x = 41$

Solve.

5. $n - 8 = 11$ — $n = 19$
6. $9 + g = 13$ — $g = 4$
7. $y + 6 = 2$ — $y = -4$
8. $-6 + j = -12$ — $j = -6$
9. $s - 8 = 11$ — $s = 19$
10. $-16 + r = -2$ — $r = 14$
11. $a + 35 = 51$ — $a = 16$
12. $m - 6 = -13$ — $m = -7$
13. $d - 12 = -5$ — $d = 7$
14. $7.5 + c = 10.6$ — $c = 3.1$
15. $y - 1.7 = 0.6$ — $y = 2.3$
16. $m - 2.25 = 4.50$ — $m = 6.75$

17. Two sisters, Jenny and Penny, play on the same basketball team. Last season they scored a combined total of 458 points. Jenny scored 192 of the points. Write and solve an equation to find the number of points Penny scored.

$192 + p = 458$; $p = 266$

18. After his payment, Mr. Weber's credit card balance was $245.76. His payment was for $75.00. Write and solve an equation to find the amount of his credit card bill.

$x - 75.00 = 245.76$; $x = 320.76$

Copyright © by Holt, Rinehart and Winston. All rights reserved.

LECCIÓN 1-7

Práctica

Cómo resolver ecuaciones mediante la suma o la resta

Determina qué valor es una solución de la ecuación.

1. $x - 6 = 12$; $x = 6, 8$ ó 18 — $x = 18$
2. $9 + x = 17$; $x = 6, 8$ ó 26 — $x = 8$
3. $x - 12 = 26$; $x = 14, 38$ ó 40 — $x = 38$
4. $x + 18 = 59$; $x = 37, 41$ ó 77 — $x = 41$

Resuelve.

5. $n - 8 = 11$ — $n = 19$
6. $9 + g = 13$ — $g = 4$
7. $y + 6 = 2$ — $y = -4$
8. $-6 + j = -12$ — $j = -6$
9. $s - 8 = 11$ — $s = 19$
10. $-16 + r = -2$ — $r = 14$
11. $a + 35 = 51$ — $a = 16$
12. $m - 6 = -13$ — $m = -7$
13. $d - 12 = -5$ — $d = 7$
14. $7.5 + c = 10.6$ — $c = 3.1$
15. $y - 1.7 = 0.6$ — $y = 2.3$
16. $m - 2.25 = 4.50$ — $m = 6.75$

17. Dos hermanas, Jenny y Penny, juegan en el mismo equipo de básquetbol. En la temporada pasada, anotaron un total de 458 puntos entre las dos. Jenny anotó 192 puntos. Escribe y resuelve una ecuación para hallar la cantidad de puntos que anotó Penny.

$192 + p = 458$; $p = 266$

18. Después de hacer un pago, el saldo de la tarjeta de crédito del señor Weber era de $245.76. El pago fue de $75. Escribe y resuelve una ecuación para hallar de cuánto era la cuenta de su tarjeta de crédito.

$x - 75 = 245.76$; $x = 320.76$

Copyright © by Holt, Rinehart and Winston. All rights reserved.

LESSON 1-8

Practice

Solving Equations by Multiplying or Dividing

Solve and check.

1. $4w = 48$ — $w = 12$
2. $8y = 56$ — $y = 7$
3. $-4b = 64$ — $b = -16$
4. $\frac{x}{4} = -9$ — $x = -36$
5. $\frac{v}{-6} = -14$ — $v = 84$
6. $\frac{n}{21} = -3$ — $n = -63$
7. $5a = -75$ — $a = -15$
8. $54 = 3q$ — $q = 18$
9. $23b = 161$ — $b = 7$
10. $\frac{k}{21} = 15$ — $k = 315$
11. $\frac{w}{-17} = 17$ — $w = -289$
12. $11 = \frac{r}{34}$ — $r = 374$
13. $672 = -24b$ — $b = -28$
14. $\frac{u}{25} = 13$ — $u = 325$
15. $42m = -966$ — $m = -23$
16. $3x + 7 = 16$ — $x = 3$
17. $\frac{t}{5} + 8 = 10$ — $t = 10$
18. $5 = 2n - 3$ — $n = 4$

19. Alex scored 13 points in the basketball game. This was $\frac{1}{5}$ of the total points the team scored. Write and solve an equation to determine the total points t the team scored.

$\frac{t}{5} = 13$; $t = 65$

20. Jar candles at the Candle Co. cost $4. Nikki spent $92 buying jar candles for party favors. Write and solve an equation to determine how many jar candles c Nikki bought at the Candle Co.

$4c = 92$; $c = 23$

Copyright © by Holt, Rinehart and Winston. All rights reserved.

LECCIÓN 1-8

Práctica

Cómo resolver ecuaciones mediante la multiplicación o la división

Resuelve y comprueba.

1. $4w = 48$ — $w = 12$
2. $8y = 56$ — $y = 7$
3. $-4b = 64$ — $b = -16$
4. $\frac{x}{4} = -9$ — $x = -36$
5. $\frac{v}{-6} = -14$ — $v = 84$
6. $\frac{n}{21} = -3$ — $n = -63$
7. $5a = -75$ — $a = -15$
8. $54 = 3q$ — $q = 18$
9. $23b = 161$ — $b = 7$
10. $\frac{k}{21} = 15$ — $k = 315$
11. $\frac{w}{-17} = 17$ — $w = -289$
12. $11 = \frac{r}{34}$ — $r = 374$
13. $672 = -24b$ — $b = -28$
14. $\frac{u}{25} = 13$ — $u = 325$
15. $42m = -966$ — $m = -23$
16. $3x + 7 = 16$ — $x = 3$
17. $\frac{t}{5} + 8 = 10$ — $t = 10$
18. $5 = 2n - 3$ — $n = 4$

19. Alex anotó 13 puntos en un juego de básquetbol. Esto representa $\frac{1}{5}$ del total de puntos que anotó su equipo. Escribe y resuelve una ecuación para determinar el total de puntos e que anotó su equipo.

$\frac{t}{5} = 13$; $e = 65$

20. En la Compañía de Velas, los vasos con velas cuestan $4. Nikki gastó $92 en vasos con velas para regalar en una fiesta. Escribe y resuelve una ecuación para determinar cuántos vasos con velas, v, compró Nikki en la Compañía de Velas.

$4c = 92$; $v = 23$

Copyright © by Holt, Rinehart and Winston. All rights reserved.

Copyright © by Holt, Rinehart and Winston. All rights reserved.

LESSON **1-9**

Practice

Introduction to Inequalities

Compare each inequality. Write < or >.

1. 7 + 10 [>] 16
2. 21 [>] 4(5)
3. 25 − 7 [<] 19
4. 58 [>] 7(8)
5. −4(8) [<] −30
6. 3 − 8 [<] −2
7. 7 + (−7) [>] −17
8. 9(−7) [>] −70
9. −43 + (−18) [<] −23

Solve and graph each inequality.

10. $x + 4 > 9$ — 0 1 2 3 4 5 6 — $x > 5$
11. $c - 6 \le 1$ — 3 4 5 6 7 8 9 — $c \le 7$
12. $y + 3 \ge -8$ — −13 −12 −11 −10 −9 −8 −7 — $y \ge -11$
13. $3 + v < -5$ — −11 −10 −9 −8 −7 −6 −5 — $v < -8$
14. $7 + x \le 10$ — 0 1 2 3 4 5 6 — $x \le 3$
15. $s - 4 < -10$ — −9 −8 −7 −6 −5 −4 −3 — $s < -6$
16. $b - 2 \le 5$ — 3 4 5 6 7 8 9 — $b \le 7$
17. $7 + n > -2$ — −11 −10 −9 −8 −7 −6 −5 — $n > -9$
18. $r + 6 \ge -1$ — −9 −8 −7 −6 −5 −4 −3 — $r \ge -7$
19. $-9 + w < -15$ — −9 −8 −7 −6 −5 −4 −3 — $w < -6$
20. $14 + k > 25$ — 8 9 10 11 12 13 14 — $k > 11$
21. $a - 8 \ge -12$ — −6 −5 −4 −3 −2 −1 0 — $a \ge -4$
22. $k + 3 \le 0$ — −5 −4 −3 −2 −1 0 1 — $k \le -3$
23. $n + 6 \ge 2$ — −6 −5 −4 −3 −2 −1 0 — $n \ge -4$
24. $-1 + b \le -1$ — −3 −2 −1 0 1 2 3 — $b \le 0$

Copyright © by Holt, Rinehart and Winston. All rights reserved.

LECCIÓN **1-9**

Práctica

Introducción a las desigualdades

Compara cada desigualdad. Escribe < ó >.

1. 7 + 10 [>] 16
2. 21 [>] 4(5)
3. 25 − 7 [<] 19
4. 58 [>] 7(8)
5. −4(8) [<] −30
6. 3 − 8 [<] −2
7. 7 + (−7) [>] −17
8. 9(−7) [>] −70
9. −43 + (−18) [<] −23

Resuelve y representa gráficamente cada desigualdad.

10. $x + 4 > 9$ — 0 1 2 3 4 5 6 — $x > 5$
11. $c - 6 \le 1$ — 3 4 5 6 7 8 9 — $c \le 7$
12. $y + 3 \ge -8$ — −13 −12 −11 −10 −9 −8 −7 — $y \ge -11$
13. $3 + v < -5$ — −11 −10 −9 −8 −7 −6 −5 — $v < -8$
14. $7 + x \le 10$ — 0 1 2 3 4 5 6 — $x \le 3$
15. $s - 4 < -10$ — −9 −8 −7 −6 −5 −4 −3 — $s < -6$
16. $b - 2 \le 5$ — 3 4 5 6 7 8 9 — $b \le 7$
17. $7 + n > -2$ — −11 −10 −9 −8 −7 −6 −5 — $n > -9$
18. $r + 6 \ge -1$ — −9 −8 −7 −6 −5 −4 −3 — $r \ge -7$
19. $-9 + w < -15$ — −9 −8 −7 −6 −5 −4 −3 — $w < -6$
20. $14 + k > 25$ — 8 9 10 11 12 13 14 — $k > 11$
21. $a - 8 \ge -12$ — −6 −5 −4 −3 −2 −1 0 — $a \ge -4$
22. $k + 3 \le 0$ — −5 −4 −3 −2 −1 0 1 — $k \le -3$
23. $n + 6 \ge 2$ — −6 −5 −4 −3 −2 −1 0 — $n \ge -4$
24. $-1 + b \le -1$ — −3 −2 −1 0 1 2 3 — $b \le 0$

Copyright © by Holt, Rinehart and Winston. All rights reserved.

LESSON **2-1**

Practice

Rational Numbers

Simplify.

1. $\frac{6}{9}$ — $\frac{2}{3}$
2. $\frac{48}{96}$ — $\frac{1}{2}$
3. $\frac{13}{52}$ — $\frac{1}{4}$
4. $-\frac{7}{28}$ — $-\frac{1}{4}$
5. $\frac{15}{40}$ — $\frac{3}{8}$
6. $-\frac{4}{48}$ — $-\frac{1}{12}$
7. $-\frac{14}{63}$ — $-\frac{2}{9}$
8. $\frac{12}{72}$ — $\frac{1}{6}$

Write each decimal as a fraction in simplest form.

9. 0.72 — $\frac{18}{25}$
10. 0.058 — $\frac{29}{500}$
11. −1.65 — $-1\frac{13}{20}$
12. 2.1 — $2\frac{1}{10}$
13. 0.036 — $\frac{9}{250}$
14. −4.06 — $-4\frac{3}{50}$
15. 2.305 — $2\frac{61}{200}$
16. 0.0064 — $\frac{4}{625}$
17. −0.60 — $-\frac{3}{5}$
18. 6.95 — $6\frac{19}{20}$
19. 0.016 — $\frac{2}{125}$
20. 0.0005 — $\frac{1}{2000}$

Write each fraction as a decimal.

21. $\frac{1}{8}$ — 0.125
22. $\frac{8}{3}$ — $2.66\overline{6}$
23. $\frac{14}{15}$ — $0.93\overline{3}$
24. $\frac{16}{5}$ — 3.2
25. $\frac{11}{16}$ — 0.6875
26. $\frac{7}{9}$ — $0.77\overline{7}$
27. $\frac{4}{5}$ — 0.8
28. $\frac{31}{25}$ — 1.24

29. Make up a fraction that cannot be simplified that has 24 as its denominator.
sample answer: $\frac{5}{24}$

Copyright © by Holt, Rinehart and Winston. All rights reserved.

LECCIÓN **2-1**

Práctica

Números racionales

Simplifica.

1. $\frac{6}{9}$ — $\frac{2}{3}$
2. $\frac{48}{96}$ — $\frac{1}{2}$
3. $\frac{13}{52}$ — $\frac{1}{4}$
4. $-\frac{7}{28}$ — $-\frac{1}{4}$
5. $\frac{15}{40}$ — $\frac{3}{8}$
6. $-\frac{4}{48}$ — $-\frac{1}{12}$
7. $-\frac{14}{63}$ — $-\frac{2}{9}$
8. $\frac{12}{72}$ — $\frac{1}{6}$

Escribe cada decimal como una fracción en su mínima expresión.

9. 0.72 — $\frac{18}{25}$
10. 0.058 — $\frac{29}{500}$
11. −1.65 — $-1\frac{13}{20}$
12. 2.1 — $2\frac{1}{10}$
13. 0.036 — $\frac{9}{250}$
14. −4.06 — $-4\frac{3}{50}$
15. 2.305 — $2\frac{61}{200}$
16. 0.0064 — $\frac{4}{625}$
17. −0.60 — $-\frac{3}{5}$
18. 6.95 — $6\frac{19}{20}$
19. 0.016 — $\frac{2}{125}$
20. 0.0005 — $\frac{1}{2000}$

Escribe cada fracción como un decimal.

21. $\frac{1}{8}$ — 0.125
22. $\frac{8}{3}$ — $2.66\overline{6}$
23. $\frac{14}{15}$ — $0.93\overline{3}$
24. $\frac{16}{5}$ — 3.2
25. $\frac{11}{16}$ — 0.6875
26. $\frac{7}{9}$ — $0.77\overline{7}$
27. $\frac{4}{5}$ — 0.8
28. $\frac{31}{25}$ — 1.24

29. Escribe una fracción que no se pueda simplificar y que tenga 24 como denominador.
ejemplo de respuesta: $\frac{5}{24}$

Copyright © by Holt, Rinehart and Winston. All rights reserved.

Copyright © by Holt, Rinehart and Winston. All rights reserved.

LESSON **2-2**
Practice
Comparing and Ordering Rational Numbers

Compare. Write <, >, or =.

1. $\frac{1}{8} \boxed{>} \frac{1}{10}$
2. $\frac{3}{5} \boxed{<} \frac{7}{10}$
3. $-\frac{1}{3} \boxed{>} -\frac{3}{4}$
4. $\frac{5}{6} \boxed{>} \frac{3}{4}$
5. $-\frac{2}{7} \boxed{>} -\frac{1}{2}$
6. $1\frac{2}{9} \boxed{<} 1\frac{2}{3}$
7. $-\frac{8}{9} \boxed{<} -\frac{3}{10}$
8. $-\frac{4}{5} \boxed{=} -\frac{8}{10}$
9. $0.08 \boxed{<} \frac{3}{10}$
10. $\frac{11}{15} \boxed{=} 0.7\overline{3}$
11. $2\frac{4}{9} \boxed{<} 2\frac{3}{4}$
12. $-\frac{5}{8} \boxed{<} -0.58$
13. $3\frac{1}{4} \boxed{<} 3.3$
14. $-\frac{1}{6} \boxed{>} -\frac{1}{9}$
15. $0.75 \boxed{=} \frac{3}{4}$
16. $-2\frac{1}{8} \boxed{<} -2.1$
17. $1\frac{1}{2} \boxed{>} 1.456$
18. $-\frac{3}{5} \boxed{=} -0.6$

19. On Monday, Gina ran 1 mile in 9.3 minutes. Her times for running 1 mile on each of the next four days, relative to her time on Monday, were $-1\frac{2}{3}$ minutes, -1.45 minutes, -1.8 minutes, and $-1\frac{3}{8}$ minutes. List these relative times in order from least to greatest.

-1.8 minutes, $-1\frac{2}{3}$ minutes, -1.45 minutes, $-1\frac{3}{8}$ minutes

20. Trail A is 3.1 miles long. Trail C is $3\frac{1}{4}$ miles long. Trail B is longer than Trail A but shorter than Trail C. What is a reasonable distance for the length of Trail B?

Possible answer: 3.2 miles

Copyright © by Holt, Rinehart and Winston.
All rights reserved.

LECCIÓN **2-2**
Práctica
Cómo comparar y ordenar números racionales

Compara. Escribe <, > ó =.

1. $\frac{1}{8} \boxed{>} \frac{1}{10}$
2. $\frac{3}{5} \boxed{<} \frac{7}{10}$
3. $-\frac{1}{3} \boxed{>} -\frac{3}{4}$
4. $\frac{5}{6} \boxed{>} \frac{3}{4}$
5. $-\frac{2}{7} \boxed{>} -\frac{1}{2}$
6. $1\frac{2}{9} \boxed{<} 1\frac{2}{3}$
7. $-\frac{8}{9} \boxed{<} -\frac{3}{10}$
8. $-\frac{4}{5} \boxed{=} -\frac{8}{10}$
9. $0.08 \boxed{<} \frac{3}{10}$
10. $\frac{11}{15} \boxed{=} 0.7\overline{3}$
11. $2\frac{4}{9} \boxed{<} 2\frac{3}{4}$
12. $-\frac{5}{8} \boxed{<} -0.58$
13. $3\frac{1}{4} \boxed{<} 3.3$
14. $-\frac{1}{6} \boxed{>} -\frac{1}{9}$
15. $0.75 \boxed{=} \frac{3}{4}$
16. $-2\frac{1}{8} \boxed{<} -2.1$
17. $1\frac{1}{2} \boxed{>} 1.456$
18. $-\frac{3}{5} \boxed{=} -0.6$

19. El lunes, Gina corrió 1 milla en 9.3 minutos. El tiempo que le llevó correr 1 milla los cuatro días siguientes en relación con el tiempo que le llevó el lunes fue de $-1\frac{2}{3}$ minutos, -1.45 minutos, -1.8 minutos y $-1\frac{3}{8}$ minutos. Ordena estos tiempos relativos de menor a mayor.

-1.8 minutos, $-1\frac{2}{3}$ minutos, -1.45 minutos, $-1\frac{3}{8}$ minutos

20. El sendero A mide 3.1 millas de longitud. El sendero C mide $3\frac{1}{4}$ millas de longitud. El sendero B es más largo que el sendero A, pero más corto que el sendero C. ¿Cuál es una distancia razonable para representar la longitud del sendero B?

Respuesta posible: 3.2 millas

Copyright © by Holt, Rinehart and Winston.
All rights reserved.

LESSON **2-3**
Practice
Adding and Subtracting Rational Numbers

1. Gretchen bought a sweater for \$23.89. In addition, she had to pay \$1.43 in sales tax. She gave the sales clerk \$30. How much change did Gretchen receive from her total purchase?

\$4.68

2. Jacob is replacing the molding around two sides of a picture frame. The measurements of the sides of the frame are $4\frac{3}{16}$ in. and $2\frac{5}{16}$ in. What length of molding will Jacob need?

$6\frac{1}{2}$ in.

Use a number line to find each sum.

3. $-0.5 + 0.4$

-0.1

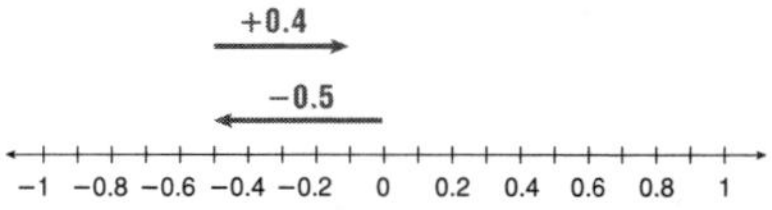

4. $-\frac{2}{7} + \frac{6}{7}$

$\frac{4}{7}$

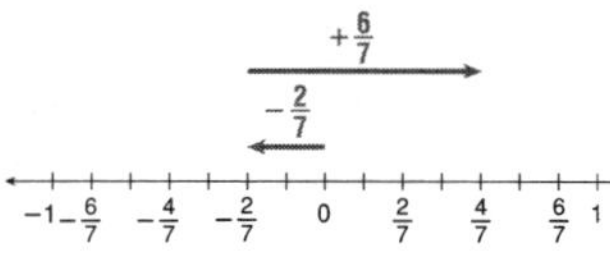

Add or subtract. Simplify.

5. $\frac{3}{8} + \frac{1}{8}$ **$\frac{1}{2}$**
6. $-\frac{1}{10} + \frac{7}{10}$ **$\frac{3}{5}$**
7. $\frac{5}{14} - \frac{3}{14}$ **$\frac{1}{7}$**
8. $\frac{4}{15} + \frac{7}{15}$ **$\frac{11}{15}$**
9. $\frac{5}{18} - \frac{7}{18}$ **$-\frac{1}{9}$**
10. $-\frac{8}{17} - \frac{2}{17}$ **$-\frac{10}{17}$**
11. $-\frac{1}{16} + \frac{5}{16}$ **$\frac{1}{4}$**
12. $\frac{3}{20} + \frac{1}{20}$ **$\frac{1}{5}$**

Evaluate each expression for the given value of the variable.

13. $38.1 + x$ for $x = -6.1$ **32**
14. $18.7 + x$ for $x = 8.5$ **27.2**
15. $\frac{8}{15} + x$ for $x = -\frac{4}{15}$ **$\frac{4}{15}$**

Copyright © by Holt, Rinehart and Winston.
All rights reserved.

LECCIÓN **2-3**
Práctica
Cómo sumar y restar números racionales

1. Gretchen compró un suéter a \$23.89. Además, tuvo que pagar \$1.43 de impuestos sobre la venta. Le dio \$30 a la vendedora. ¿Cuánto vuelto recibió Gretchen por la compra total?

\$4.68

2. Jacob está reemplazando la moldura de dos lados de un portarretratos. Las medidas de los lados del portarretratos son $4\frac{3}{16}$ pulg y $2\frac{5}{16}$ pulg. ¿Cuánta moldura necesitará?

$6\frac{1}{2}$ pulg

Usa una recta numérica para hallar cada suma.

3. $-0.5 + 0.4$

-0.1

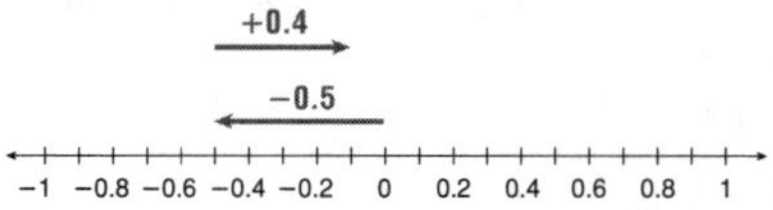

4. $-\frac{2}{7} + \frac{6}{7}$

$\frac{4}{7}$

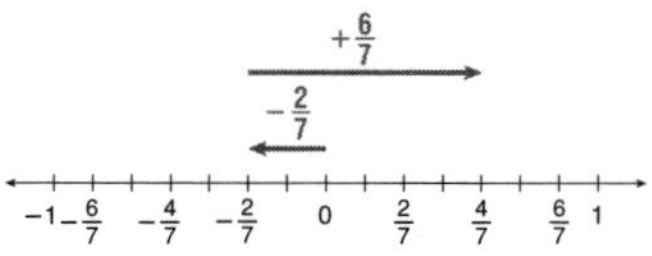

Suma o resta. Simplifica.

5. $\frac{3}{8} + \frac{1}{8}$ **$\frac{1}{2}$**
6. $-\frac{1}{10} + \frac{7}{10}$ **$\frac{3}{5}$**
7. $\frac{5}{14} - \frac{3}{14}$ **$\frac{1}{7}$**
8. $\frac{4}{15} + \frac{7}{15}$ **$\frac{11}{15}$**
9. $\frac{5}{18} - \frac{7}{18}$ **$-\frac{1}{9}$**
10. $-\frac{8}{17} - \frac{2}{17}$ **$-\frac{10}{17}$**
11. $-\frac{1}{16} + \frac{5}{16}$ **$\frac{1}{4}$**
12. $\frac{3}{20} + \frac{1}{20}$ **$\frac{1}{5}$**

Evalúa cada expresión para el valor dado de la variable.

13. $38.1 + x$ para $x = -6.1$ **32**
14. $18.7 + x$ para $x = 8.5$ **27.2**
15. $\frac{8}{15} + x$ para $x = -\frac{4}{15}$ **$\frac{4}{15}$**

Copyright © by Holt, Rinehart and Winston.
All rights reserved.

Copyright © by Holt, Rinehart and Winston.
All rights reserved.

LESSON **2-4**

Practice

Multiplying Rational Numbers

Multiply. Write each answer in simplest form.

1. $8\left(\frac{3}{4}\right)$	6
2. $-6\left(\frac{9}{18}\right)$	−3
3. $-9\left(\frac{5}{6}\right)$	$-7\frac{1}{2}$
4. $-6\left(-\frac{7}{12}\right)$	$3\frac{1}{2}$
5. $-\frac{5}{18}\left(\frac{8}{15}\right)$	$-\frac{4}{27}$
6. $\frac{7}{12}\left(\frac{14}{21}\right)$	$\frac{7}{18}$
7. $-\frac{1}{9}\left(\frac{27}{24}\right)$	$-\frac{1}{8}$
8. $-\frac{1}{11}\left(-\frac{3}{2}\right)$	$\frac{3}{22}$
9. $\frac{7}{20}\left(-\frac{15}{28}\right)$	$-\frac{3}{16}$
10. $\frac{16}{25}\left(-\frac{18}{32}\right)$	$-\frac{9}{25}$
11. $\frac{1}{9}\left(-\frac{18}{17}\right)$	$-\frac{2}{17}$
12. $\frac{17}{20}\left(-\frac{12}{34}\right)$	$-\frac{3}{10}$
13. $-4\left(2\frac{1}{6}\right)$	$-8\frac{2}{3}$
14. $\frac{3}{4}\left(1\frac{3}{8}\right)$	$1\frac{1}{32}$
15. $3\frac{1}{5}\left(\frac{2}{3}\right)$	$2\frac{2}{15}$
16. $-\frac{5}{6}\left(2\frac{1}{2}\right)$	$-2\frac{1}{12}$

Multiply.

Problem	Answer
17. −2(−5.2)	10.4
18. 0.53(0.04)	0.0212
19. (−7)(−3.9)	27.3
20. −2(8.13)	−16.26
21. 0.02(−4.62)	−0.0924
22. 0.5(−7.8)	−3.9
23. (−0.41)(−8.5)	3.485
24. (−8)(6.3)	−50.4
25. 15(−0.05)	−0.75
26. (−3.04)(−1.7)	5.168
27. 10(−0.09)	−0.9
28. (−0.8)(−0.15)	0.12

29. Travis painted for $6\frac{2}{3}$ hours. He received $27 an hour for his work. How much was Travis paid for doing this painting job?

$180

Copyright © by Holt, Rinehart and Winston. All rights reserved.

LECCIÓN **2-4**

Práctica

Cómo multiplicar números racionales

Multiplica. Escribe cada respuesta en su mínima expresión.

1. $8\left(\frac{3}{4}\right)$	6
2. $-6\left(\frac{9}{18}\right)$	−3
3. $-9\left(\frac{5}{6}\right)$	$-7\frac{1}{2}$
4. $-6\left(-\frac{7}{12}\right)$	$3\frac{1}{2}$
5. $-\frac{5}{18}\left(\frac{8}{15}\right)$	$-\frac{4}{27}$
6. $\frac{7}{12}\left(\frac{14}{21}\right)$	$\frac{7}{18}$
7. $-\frac{1}{9}\left(\frac{27}{24}\right)$	$-\frac{1}{8}$
8. $-\frac{1}{11}\left(-\frac{3}{2}\right)$	$\frac{3}{22}$
9. $\frac{7}{20}\left(-\frac{15}{28}\right)$	$-\frac{3}{16}$
10. $\frac{16}{25}\left(-\frac{18}{32}\right)$	$-\frac{9}{25}$
11. $\frac{1}{9}\left(-\frac{18}{17}\right)$	$-\frac{2}{17}$
12. $\frac{17}{20}\left(-\frac{12}{34}\right)$	$-\frac{3}{10}$
13. $-4\left(2\frac{1}{6}\right)$	$-8\frac{2}{3}$
14. $\frac{3}{4}\left(1\frac{3}{8}\right)$	$1\frac{1}{32}$
15. $3\frac{1}{5}\left(\frac{2}{3}\right)$	$2\frac{2}{15}$
16. $-\frac{5}{6}\left(2\frac{1}{2}\right)$	$-2\frac{1}{12}$

Multiplica.

Problema	Respuesta
17. −2(−5.2)	10.4
18. 0.53(0.04)	0.0212
19. (−7)(−3.9)	27.3
20. −2(8.13)	−16.26
21. 0.02(−4.62)	−0.0924
22. 0.5(−7.8)	−3.9
23. (−0.41)(−8.5)	3.485
24. (−8)(6.3)	−50.4
25. 15(−0.05)	−0.75
26. (−3.04)(−1.7)	5.168
27. 10(−0.09)	−0.9
28. (−0.8)(−0.15)	0.12

29. Travis pintó durante $6\frac{2}{3}$ horas. Recibió $27 por hora por su trabajo. ¿Cuánto le pagaron en total?

$180

Copyright © by Holt, Rinehart and Winston. All rights reserved.

LESSON **2-5**

Practice

Dividing Rational Numbers

Divide. Write each answer in simplest form.

1. $\frac{1}{5} \div \frac{3}{10}$	$\frac{2}{3}$
2. $-\frac{5}{8} \div \frac{3}{4}$	$-\frac{5}{6}$
3. $\frac{1}{4} \div \frac{1}{8}$	2
4. $-\frac{2}{3} \div \frac{4}{15}$	$-2\frac{1}{2}$
5. $1\frac{2}{9} \div 1\frac{2}{3}$	$\frac{11}{15}$
6. $-\frac{7}{10} \div \left(\frac{2}{5}\right)$	$-1\frac{3}{4}$
7. $\frac{6}{11} \div \frac{3}{22}$	4
8. $\frac{4}{9} \div \left(-\frac{8}{15}\right)$	$-\frac{5}{6}$
9. $\frac{3}{8} \div -15$	$-\frac{1}{40}$
10. $-\frac{5}{6} \div 12$	$-\frac{5}{72}$
11. $6\frac{1}{2} \div 1\frac{5}{8}$	4
12. $-\frac{9}{10} \div 6$	$-\frac{3}{20}$

Divide.

13. 24.35 ÷ 0.5	48.7
14. 2.16 ÷ 0.04	54
15. 3.16 ÷ 0.02	158
16. 7.32 ÷ 0.3	24.4
17. 87.36 ÷ 0.6	145.6
18. 79.36 ÷ 0.8	99.2
19. 4.27 ÷ 0.007	610
20. 63.81 ÷ 0.9	70.9
21. 1.23 ÷ 0.003	410
22. 62.46 ÷ 0.09	694
23. 21.12 ÷ 0.4	52.8
24. 82.68 ÷ 0.06	1378

Evaluate each expression for the given value of the variable.

25. $\frac{18}{x}$ for $x = 0.12$	150
26. $\frac{10.8}{x}$ for $x = 0.03$	360
27. $\frac{9.18}{x}$ for $x = -1.2$	−7.65

28. A can of fruit contains $3\frac{1}{2}$ cups of fruit. The suggested serving size is $\frac{1}{2}$ cup. How many servings are in the can of fruit?

7 servings

Copyright © by Holt, Rinehart and Winston. All rights reserved.

LECCIÓN **2-5**

Práctica

Cómo dividir números racionales

Divide. Escribe cada respuesta en su mínima expresión.

1. $\frac{1}{5} \div \frac{3}{10}$	$\frac{2}{3}$
2. $-\frac{5}{8} \div \frac{3}{4}$	$-\frac{5}{6}$
3. $\frac{1}{4} \div \frac{1}{8}$	2
4. $-\frac{2}{3} \div \frac{4}{15}$	$-2\frac{1}{2}$
5. $1\frac{2}{9} \div 1\frac{2}{3}$	$\frac{11}{15}$
6. $-\frac{7}{10} \div \left(\frac{2}{5}\right)$	$-1\frac{3}{4}$
7. $\frac{6}{11} \div \frac{3}{22}$	4
8. $\frac{4}{9} \div \left(-\frac{8}{15}\right)$	$-\frac{5}{6}$
9. $\frac{3}{8} \div -15$	$-\frac{1}{40}$
10. $-\frac{5}{6} \div 12$	$-\frac{5}{72}$
11. $6\frac{1}{2} \div 1\frac{5}{8}$	4
12. $-\frac{9}{10} \div 6$	$-\frac{3}{20}$

Divide.

13. 24.35 ÷ 0.5	48.7
14. 2.16 ÷ 0.04	54
15. 3.16 ÷ 0.02	158
16. 7.32 ÷ 0.3	24.4
17. 87.36 ÷ 0.6	145.6
18. 79.36 ÷ 0.8	99.2
19. 4.27 ÷ 0.007	610
20. 63.81 ÷ 0.9	70.9
21. 1.23 ÷ 0.003	410
22. 62.46 ÷ 0.09	694
23. 21.12 ÷ 0.4	52.8
24. 82.68 ÷ 0.06	1378

Evalúa cada expresión para el valor dado de la variable.

Problema	Respuesta
25. $\frac{18}{x}$ para $x = 0.12$	150
26. $\frac{10.8}{x}$ para $x = 0.03$	360
27. $\frac{9.18}{x}$ para $x = -1.2$	−7.65

28. Una lata de frutas contiene $3\frac{1}{2}$ tazas de fruta. El tamaño sugerido de la porción es de $\frac{1}{2}$ taza. ¿Cuántas porciones contiene la lata?

7 porciones

Copyright © by Holt, Rinehart and Winston. All rights reserved.

Copyright © by Holt, Rinehart and Winston.
All rights reserved.

LESSON 2-6

Practice

Adding and Subtracting with Unlike Denominators

Add or subtract.

1. $\frac{2}{3} + \frac{1}{2}$ — $1\frac{1}{6}$
2. $\frac{3}{5} + \frac{1}{3}$ — $\frac{14}{15}$
3. $\frac{3}{4} - \frac{1}{3}$ — $\frac{5}{12}$
4. $\frac{1}{2} - \frac{5}{9}$ — $-\frac{1}{18}$
5. $\frac{5}{16} - \frac{5}{8}$ — $-\frac{5}{16}$
6. $\frac{7}{9} + \frac{5}{6}$ — $1\frac{11}{18}$
7. $\frac{7}{8} - \frac{1}{4}$ — $\frac{5}{8}$
8. $\frac{5}{6} - \frac{3}{8}$ — $\frac{11}{24}$
9. $2\frac{7}{8} + 3\frac{5}{12}$ — $6\frac{7}{24}$
10. $1\frac{2}{9} + 2\frac{1}{18}$ — $3\frac{5}{18}$
11. $3\frac{2}{3} - 1\frac{3}{5}$ — $2\frac{1}{15}$
12. $1\frac{5}{6} + (-2\frac{3}{4})$ — $-\frac{11}{12}$
13. $8\frac{1}{3} - 3\frac{5}{9}$ — $4\frac{7}{9}$
14. $5\frac{1}{3} + 1\frac{11}{12}$ — $7\frac{1}{4}$
15. $7\frac{1}{4} + (-2\frac{5}{12})$ — $4\frac{5}{6}$
16. $5\frac{2}{5} - 7\frac{3}{10}$ — $-1\frac{9}{10}$

Evaluate each expression for the given value of the variable.

17. $2\frac{3}{8} + x$ for $x = 1\frac{5}{6}$ — $4\frac{5}{24}$
18. $x - \frac{2}{5}$ for $x = \frac{1}{3}$ — $-\frac{1}{15}$
19. $x - \frac{3}{10}$ for $x = \frac{3}{7}$ — $\frac{9}{70}$
20. $1\frac{5}{8} + x$ for $x = -2\frac{1}{6}$ — $-\frac{13}{24}$
21. $x - \frac{3}{4}$ for $x = \frac{1}{6}$ — $-\frac{7}{12}$
22. $x - \frac{3}{10}$ for $x = \frac{1}{2}$ — $\frac{1}{5}$
23. Ana worked $6\frac{1}{2}$ h on Monday, $5\frac{3}{4}$ h on Tuesday and $7\frac{1}{6}$ h on Friday. How many total hours did she work these three days?

    $19\frac{5}{12}$ h

LECCIÓN 2-6

Práctica

Cómo sumar y restar con denominadores distintos

Suma o resta.

1. $\frac{2}{3} + \frac{1}{2}$ — $1\frac{1}{6}$
2. $\frac{3}{5} + \frac{1}{3}$ — $\frac{14}{15}$
3. $\frac{3}{4} - \frac{1}{3}$ — $\frac{5}{12}$
4. $\frac{1}{2} - \frac{5}{9}$ — $-\frac{1}{18}$
5. $\frac{5}{16} - \frac{5}{8}$ — $-\frac{5}{16}$
6. $\frac{7}{9} + \frac{5}{6}$ — $1\frac{11}{18}$
7. $\frac{7}{8} - \frac{1}{4}$ — $\frac{5}{8}$
8. $\frac{5}{6} - \frac{3}{8}$ — $\frac{11}{24}$
9. $2\frac{7}{8} + 3\frac{5}{12}$ — $6\frac{7}{24}$
10. $1\frac{2}{9} + 2\frac{1}{18}$ — $3\frac{5}{18}$
11. $3\frac{2}{3} - 1\frac{3}{5}$ — $2\frac{1}{15}$
12. $1\frac{5}{6} + (-2\frac{3}{4})$ — $-\frac{11}{12}$
13. $8\frac{1}{3} - 3\frac{5}{9}$ — $4\frac{7}{9}$
14. $5\frac{1}{3} + 1\frac{11}{12}$ — $7\frac{1}{4}$
15. $7\frac{1}{4} + (-2\frac{5}{12})$ — $4\frac{5}{6}$
16. $5\frac{2}{5} - 7\frac{3}{10}$ — $-1\frac{9}{10}$

Evalúa cada expresión para el valor dado de la variable.

17. $2\frac{3}{8} + x$ para $x = 1\frac{5}{6}$ — $4\frac{5}{24}$
18. $x - \frac{2}{5}$ para $x = \frac{1}{3}$ — $-\frac{1}{15}$
19. $x - \frac{3}{10}$ para $x = \frac{3}{7}$ — $\frac{9}{70}$
20. $1\frac{5}{8} + x$ para $x = -2\frac{1}{6}$ — $-\frac{13}{24}$
21. $x - \frac{3}{4}$ para $x = \frac{1}{6}$ — $-\frac{7}{12}$
22. $x - \frac{3}{10}$ para $x = \frac{1}{2}$ — $\frac{1}{5}$
23. Ana trabajó $6\frac{1}{2}$ h el lunes, $5\frac{3}{4}$ h el martes y $7\frac{1}{6}$ h el viernes. ¿Cuántas horas trabajó en total los tres días?

    $19\frac{5}{12}$ h

LESSON 2-7

Practice

Solving Equations with Rational Numbers

Solve.

1. $x + 6.8 = 12.19$ — $x = 5.39$
2. $y - 10.24 = 5.3$ — $y = 15.54$
3. $0.05w = 6.25$ — $w = 125$
4. $\frac{a}{9.05} = 8.2$ — $a = 74.21$
5. $-12.41 + x = -0.06$ — $x = 12.35$
6. $\frac{d}{-8.4} = -10.2$ — $d = 85.68$
7. $-2.89 = 1.7m$ — $m = -1.7$
8. $n - 8.09 = -11.65$ — $n = -3.56$
9. $\frac{x}{5.4} = -7.18$ — $x = -38.772$
10. $\frac{7}{9} + x = 1\frac{1}{9}$ — $x = \frac{1}{3}$
11. $\frac{6}{11}y = -\frac{18}{22}$ — $y = -1\frac{1}{2}$
12. $\frac{7}{10}d = \frac{21}{20}$ — $d = 1\frac{1}{2}$
13. $x - \left(-\frac{9}{14}\right) = \frac{5}{7}$ — $x = \frac{1}{14}$
14. $x - \frac{15}{21} = 2\frac{6}{7}$ — $x = 3\frac{4}{7}$
15. $-\frac{8}{15}a = \frac{9}{10}$ — $a = -1\frac{11}{16}$
16. A recipe calls for $2\frac{1}{3}$ cups of flour and $1\frac{1}{4}$ cups of sugar. If the recipe is tripled, how much flour and sugar will be needed?

    7 cups of flour and $3\frac{3}{4}$ cups of sugar
17. Daniel filled the gas tank in his car with 14.6 gal of gas. He then drove 284.7 mi before needing to fill up his tank with gas again. How many miles did the car get to a gallon of gasoline?

    19.5 mi

LECCIÓN 2-7

Práctica

Cómo resolver ecuaciones con números racionales

Resuelve.

1. $x + 6.8 = 12.19$ — $x = 5.39$
2. $y - 10.24 = 5.3$ — $y = 15.54$
3. $0.05w = 6.25$ — $w = 125$
4. $\frac{a}{9.05} = 8.2$ — $a = 74.21$
5. $-12.41 + x = -0.06$ — $x = 12.35$
6. $\frac{d}{-8.4} = -10.2$ — $d = 85.68$
7. $-2.89 = 1.7m$ — $m = -1.7$
8. $n - 8.09 = -11.65$ — $n = -3.56$
9. $\frac{x}{5.4} = -7.18$ — $x = -38.772$
10. $\frac{7}{9} + x = 1\frac{1}{9}$ — $x = \frac{1}{3}$
11. $\frac{6}{11}y = -\frac{18}{22}$ — $y = -1\frac{1}{2}$
12. $\frac{7}{10}d = \frac{21}{20}$ — $d = 1\frac{1}{2}$
13. $x - \left(-\frac{9}{14}\right) = \frac{5}{7}$ — $x = \frac{1}{14}$
14. $x - \frac{15}{21} = 2\frac{6}{7}$ — $x = 3\frac{4}{7}$
15. $-\frac{8}{15}a = \frac{9}{10}$ — $a = -1\frac{11}{16}$
16. Para preparar una receta se necesitan $2\frac{1}{3}$ tazas de harina y $1\frac{1}{4}$ tazas de azúcar. Si se triplica la receta, ¿cuánta harina y azúcar se necesitará?

    7 tazas de harina y $3\frac{3}{4}$ tazas de azúcar
17. Daniel llenó el tanque de su automóvil con 14.6 galones de gasolina. Luego, condujo 284.7 millas antes de tener que llenar el tanque otra vez. ¿Cuántas millas puede recorrer el automóvil con un galón de gasolina?

    19.5 millas

LESSON **2-8**

Practice
Solving Two-Step Equations

Write and solve a two-step equation to answer the following questions.

1. The school purchased baseball equipment and uniforms for a total cost of $1762. The equipment costs $598 and the uniforms were $24.25 each. How many uniforms did the school purchase?

   x = # of uniforms

   $1762 = 598 + 24.25x$

   $1762 - 598$

   $= 598 - 598 + 24.25x$

   $1164 = 24.25x$

   $\frac{1164}{24.25} = \frac{24.25x}{24.25}$

   $48 = x$

2. Carla runs 4 miles every day. She jogs from home to the school track, which is $\frac{3}{4}$ mile away. She then runs laps around the $\frac{1}{4}$-mile track. Carla then jogs home. How many laps does she run at the school?

   x = # of laps

   $4 = \frac{3}{4} + \frac{1}{4}x + \frac{3}{4}$

   $4 - \frac{6}{4} = \frac{6}{4} - \frac{6}{4} + \frac{1}{4}x$

   $\frac{5}{2} = \frac{1}{4}x$

   $4\left(\frac{5}{2}\right) = \left(\frac{1}{4}x\right)4$

   $10 = x$

Solve.

3. $\frac{a+5}{3} = 12$ — $a = 31$
4. $\frac{x+2}{4} = -2$ — $x = -10$
5. $\frac{y-4}{6} = -3$ — $y = -14$
6. $\frac{k+1}{8} = 7$ — $k = 55$
7. $0.5x - 6 = -4$ — $x = 4$
8. $\frac{x}{2} + 3 = -4$ — $x = -14$
9. $\frac{1}{5}n + 3 = 6$ — $n = 15$
10. $2a - 7 = -9$ — $a = -1$
11. $\frac{3x-1}{4} = 2$ — $x = 3$
12. $-7.8 = 4.4 + 2r$ — $r = -6.1$
13. $\frac{-4w+5}{-3} = -7$ — $w = -4$
14. $1.3 - 5r = 7.4$ — $r = -1.22$

15. A phone call costs $0.58 for the first 3 minutes and $0.15 for each additional minute. If the total charge for the call was $4.78, how many minutes was the call? **31 minutes**

16. Seventeen less than four times a number is twenty-seven. Find the number. **11**

Copyright © by Holt, Rinehart and Winston. All rights reserved.

LECCIÓN **2-8**

Práctica
Cómo resolver ecuaciones de dos pasos

Escribe y resuelve una ecuación de dos pasos para responder a las siguientes preguntas.

1. La escuela compró equipos y uniformes de béisbol por un costo total de $1762. El equipo costó $598 y los uniformes, $24.25 cada uno. ¿Cuántos uniformes compró la escuela?

   x = cantidad de uniformes

   $1762 = 598 + 24.25x$

   $1762 - 598$

   $= 598 - 598 + 24.25x$

   $1164 = 24.25x$

   $\frac{1164}{24.25} = \frac{24.25x}{24.25}$

   $48 = x$

2. Carla corre 4 millas todos los días. Corre desde su casa hasta la pista de la escuela, que está a $\frac{3}{4}$ de milla de distancia. Luego, corre unas vueltas alrededor de la pista, que mide $\frac{1}{4}$-de milla. Después, Carla corre hasta su casa. ¿Cuántas vueltas corre en la escuela?

   x = cantidad de vueltas

   $4 = \frac{3}{4} + \frac{1}{4}x + \frac{3}{4}$

   $4 - \frac{6}{4} = \frac{6}{4} - \frac{6}{4} + \frac{1}{4}x$

   $\frac{5}{2} = \frac{1}{4}x$

   $4\left(\frac{5}{2}\right) = \left(\frac{1}{4}x\right)4$

   $10 = x$

Resuelve.

3. $\frac{a+5}{3} = 12$ — $a = 31$
4. $\frac{x+2}{4} = -2$ — $x = -10$
5. $\frac{y-4}{6} = -3$ — $y = -14$
6. $\frac{k+1}{8} = 7$ — $k = 55$
7. $0.5x - 6 = -4$ — $x = 4$
8. $\frac{x}{2} + 3 = -4$ — $x = -14$
9. $\frac{1}{5}n + 3 = 6$ — $n = 15$
10. $2a - 7 = -9$ — $a = -1$
11. $\frac{3x-1}{4} = 2$ — $x = 3$
12. $-7.8 = 4.4 + 2r$ — $r = -6.1$
13. $\frac{-4w+5}{-3} = -7$ — $w = -4$
14. $1.3 - 5r = 7.4$ — $r = -1.22$

15. Una llamada telefónica cuesta $0.58 los primeros 3 minutos y $0.15 cada minuto adicional. Si el costo total de la llamada fue de $4.78, ¿cuántos minutos duró la llamada? **31 minutos**

16. Diecisiete menos que cuatro multiplicado por un número es igual a veintisiete. Halla el número. **11**

Copyright © by Holt, Rinehart and Winston. All rights reserved.

LESSON **3-1**

Practice
Ordered Pairs

Determine whether each ordered pair is a solution of $y = 4 + 2x$.

1. (1, 1) — no
2. (2, 8) — yes
3. (0, 4) — yes
4. (8, 2) — no

Determine whether each ordered pair is a solution of $y = 3x - 2$.

5. (1, 1) — yes
6. (3, 7) — yes
7. (5, 15) — no
8. (6, 16) — yes

Use the given values to complete the table of solutions.

9. $y = x + 5$ for $x = 0, 1, 2, 3, 4$

x	x + 5	y	(x, y)
0	0 + 5	5	(0, 5)
1	1 + 5	6	(1, 6)
2	2 + 5	7	(2, 7)
3	3 + 5	8	(3, 8)
4	4 + 5	9	(4, 9)

10. $y = 3x + 1$ for $x = 1, 2, 3, 4, 5$

x	3x + 1	y	(x, y)
1	3(1) + 1	4	(1, 4)
2	3(2) + 1	7	(2, 7)
3	3(3) + 1	10	(3, 10)
4	3(4) + 1	13	(4, 13)
5	3(5) + 1	16	(5, 16)

11. $y = 2x + 6$ for $x = 0, 1, 2, 3, 4$

x	2x + 6	y	(x, y)
0	2(0) + 6	6	(0, 6)
1	2(1) + 6	8	(1, 8)
2	2(2) + 6	10	(2, 10)
3	2(3) + 6	12	(3, 12)
4	2(4) + 6	14	(4, 14)

12. $y = 4x - 2$ for $x = 2, 4, 6, 8, 10$

x	4x − 2	y	(x, y)
2	4(2) − 2	6	(2, 6)
4	4(4) − 2	14	(4, 14)
6	4(6) − 2	22	(6, 22)
8	4(8) − 2	30	(8, 30)
10	4(10) − 2	38	(10, 38)

13. Alexis opened a savings account with a $120 deposit. Each week she will put $20 into the account. The equation that gives the total amount t in her account is $t = 120 + 20w$, where w is the number of weeks since she opened the account. How much money will Alexis have in her savings account after 5 weeks?

    $220

Copyright © by Holt, Rinehart and Winston. All rights reserved.

LECCIÓN **3-1**

Práctica
Pares ordenados

Determina si cada par ordenado es una solución de $y = 4 + 2x$.

1. (1, 1) — no
2. (2, 8) — sí
3. (0, 4) — sí
4. (8, 2) — no

Determina si cada par ordenado es una solución de $y = 3x - 2$.

5. (1, 1) — sí
6. (3, 7) — sí
7. (5, 15) — no
8. (6, 16) — sí

Usa los valores dados para completar la tabla de soluciones.

9. $y = x + 5$ para $x = 0, 1, 2, 3, 4$

x	x + 5	y	(x, y)
0	0 + 5	5	(0, 5)
1	1 + 5	6	(1, 6)
2	2 + 5	7	(2, 7)
3	3 + 5	8	(3, 8)
4	4 + 5	9	(4, 9)

10. $y = 3x + 1$ para $x = 1, 2, 3, 4, 5$

x	3x + 1	y	(x, y)
1	3(1) + 1	4	(1, 4)
2	3(2) + 1	7	(2, 7)
3	3(3) + 1	10	(3, 10)
4	3(4) + 1	13	(4, 13)
5	3(5) + 1	16	(5, 16)

11. $y = 2x + 6$ para $x = 0, 1, 2, 3, 4$

x	2x + 6	y	(x, y)
0	2(0) + 6	6	(0, 6)
1	2(1) + 6	8	(1, 8)
2	2(2) + 6	10	(2, 10)
3	2(3) + 6	12	(3, 12)
4	2(4) + 6	14	(4, 14)

12. $y = 4x - 2$ para $x = 2, 4, 6, 8, 10$

x	4x − 2	y	(x, y)
2	4(2) − 2	6	(2, 6)
4	4(4) − 2	14	(4, 14)
6	4(6) − 2	22	(6, 22)
8	4(8) − 2	30	(8, 30)
10	4(10) − 2	38	(10, 38)

13. Alexis abrió una cuenta de ahorros con un depósito de $120. Cada semana pondrá $20 en la cuenta. La ecuación que da la cantidad total t que tiene en la cuenta es $t = 120 + 20s$, donde s es la cantidad de semanas que pasaron desde que abrió la cuenta. ¿Cuánto dinero tendrá Alexis en su cuenta de ahorros después de 5 semanas?

    $220

Copyright © by Holt, Rinehart and Winston. All rights reserved.

Copyright © by Holt, Rinehart and Winston. All rights reserved.

LESSON 3-2
Practice
Graphing on a Coordinate Plane

Give the coordinates of each point and quadrant.

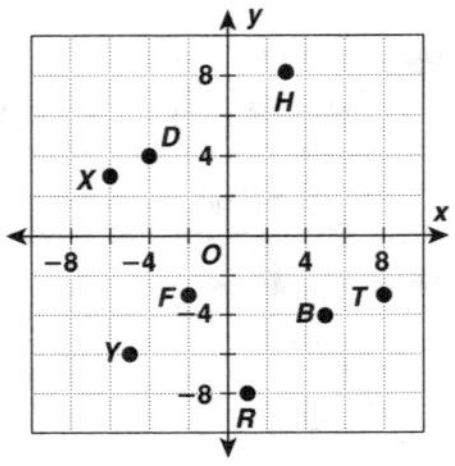

1. F (−2, −3); III
2. X (−6, 3); II
3. T (8, −3); IV
4. B (5, −4); IV
5. D (−4, 4); II
6. R (1, −8); IV
7. H (3, 8); I
8. Y (−5, −6); III

Graph each point on a coordinate plane.

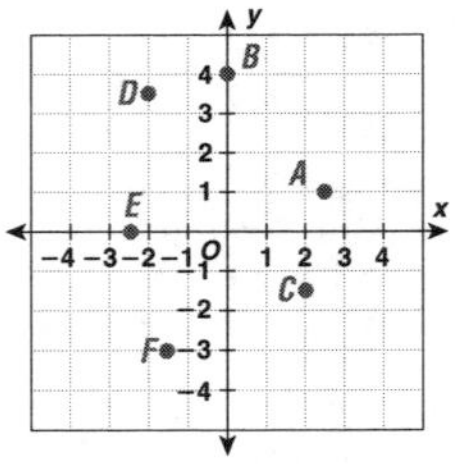

9. $A(2\frac{1}{2}, 1)$
10. $B(0, 4)$
11. $C(2, -1.5)$
12. $D(-2, 3.5)$
13. $E(-2\frac{1}{3}, 0)$
14. $F(-1\frac{1}{2}, -3)$

Complete the table of ordered pairs. Graph each ordered pair on a coordinate plane. Draw a line through the points.

15. $y = 1\frac{1}{2}x$

x	$1\frac{1}{2}x$	y	(x, y)
0	$1\frac{1}{2}(0)$	0	(0, 0)
1	$1\frac{1}{2}(1)$	$1\frac{1}{2}$	$(1, 1\frac{1}{2})$
2	$1\frac{1}{2}(2)$	3	(2, 3)

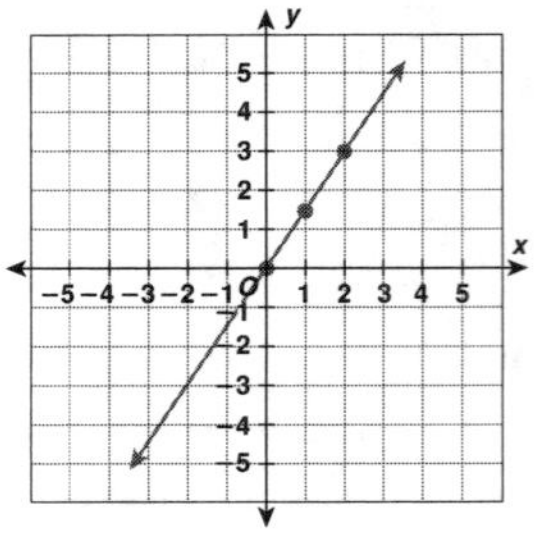

Copyright © by Holt, Rinehart and Winston. All rights reserved.

LECCIÓN 3-2
Práctica
Gráficas en un plano cartesiano

Da las coordenadas y el cuadrante de cada punto.

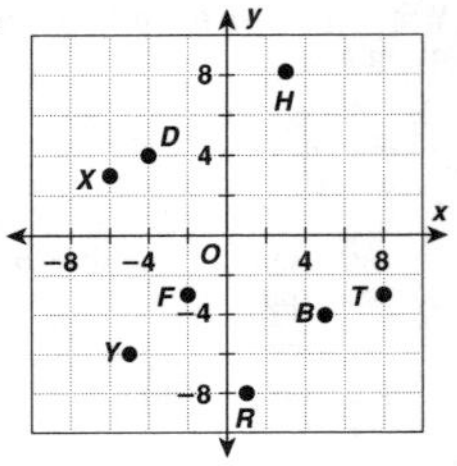

1. F (−2, −3); III
2. X (−6, 3); II
3. T (8, −3); IV
4. B (5, −4); IV
5. D (−4, 4); II
6. R (1, −8); IV
7. H (3, 8); I
8. Y (−5, −6); III

Representa gráficamente cada punto en un plano cartesiano.

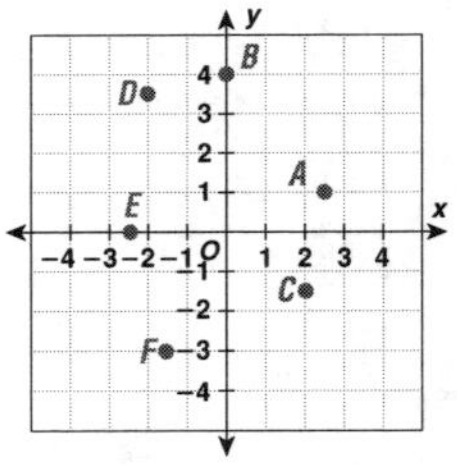

9. $A(2\frac{1}{2}, 1)$
10. $B(0, 4)$
11. $C(2, -1.5)$
12. $D(-2, 3.5)$
13. $E(-2\frac{1}{3}, 0)$
14. $F(-1\frac{1}{2}, -3)$

Completa la tabla de pares ordenados. Representa gráficamente cada par ordenado en un plano cartesiano. Dibuja una línea a través de los puntos.

15. $y = 1\frac{1}{2}x$

x	$1\frac{1}{2}x$	y	(x, y)
0	$1\frac{1}{2}(0)$	0	(0, 0)
1	$1\frac{1}{2}(1)$	$1\frac{1}{2}$	$(1, 1\frac{1}{2})$
2	$1\frac{1}{2}(2)$	3	(2, 3)

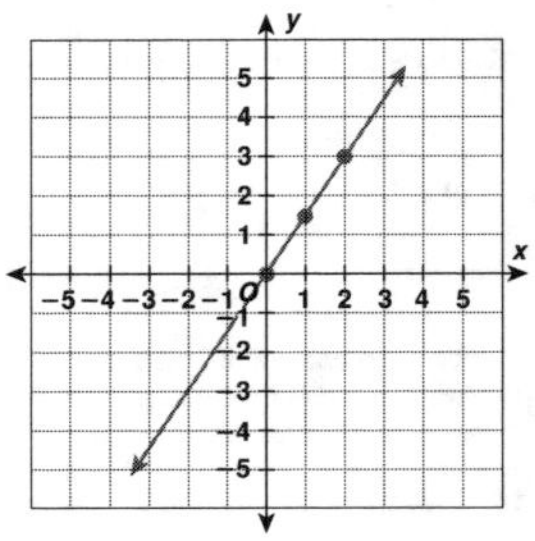

Copyright © by Holt, Rinehart and Winston. All rights reserved.

LESSON 3-3
Practice
Interpreting Graphs and Tables

The table gives the speed of three dogs in mi/h at the given times. Tell which dog corresponds to each situation described below.

Time	5:00	5:01	5:02	5:03	5:04
Dog 1	0	1	12	0	0
Dog 2	5	23	4	0	0
Dog 3	14	0	18	2	9

1. Leshaan walks his dog. Then he lets the dog off the leash and it runs around the yard. Then they go into the house and the dog stands eating from his dog dish and drinking from his water bowl. Dog 2
2. Luke's dog is chasing its tail. Then it stops and pants. The dog then runs to the backyard fence and walks along the fence, barking at a neighbor. Then it runs to Luke at the back door. Dog 3

Tell which graph corresponds to each situation in Exercises 1–2.

3.

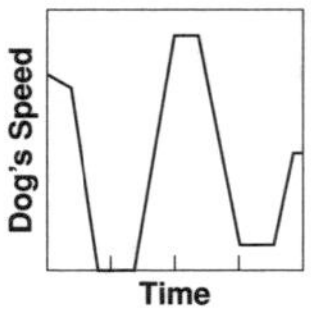

Exercise 2 (Dog 3)

4.

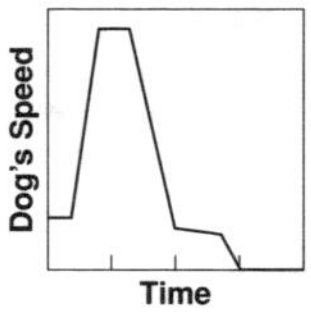

Exercise 1 (Dog 2)

5. Create a graph that illustrates the temperature inside the car.

Location	Temperature on Arrival	Temperature on Departure
Home	—	74° at 8:30
Summer job	77° at 9:00	128° at 12:05
Pool	92° at 12:15	136° at 2:30
Library	95° at 2:40	77° at 5:10

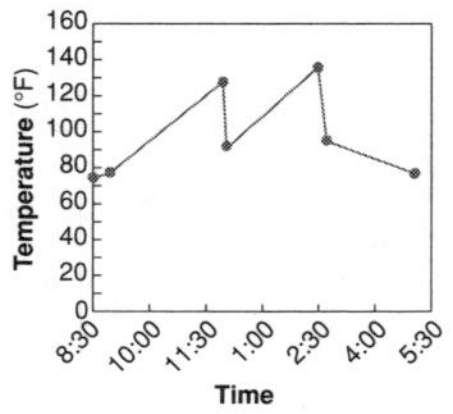

Copyright © by Holt, Rinehart and Winston. All rights reserved.

LECCIÓN 3-3
Práctica
Cómo interpretar gráficas y tablas

En la tabla se dan las velocidades de tres perros en mi/h en momentos determinados. Indica qué perro corresponde a cada situación descrita a continuación.

Hora	5:00	5:01	5:02	5:03	5:04
Perro 1	0	1	12	0	0
Perro 2	5	23	4	0	0
Perro 3	14	0	18	2	9

1. Leshaan pasea a su perro. Luego, le quita la correa y el perro corre por el jardín. Después, entran a la casa y el perro se queda quieto, comiendo y bebiendo de su plato. Perro 2
2. El perro de Luke está persiguiendo su cola. Luego, se detiene y jadea. Después, el perro corre hasta la cerca del jardín y camina a lo largo de la cerca mientras le ladra a un vecino. Luego, corre hacia Luke, que está en la puerta trasera. Perro 3

Indica que gráfica corresponde a cada situación descrita en los ejercicios 1-2.

3.

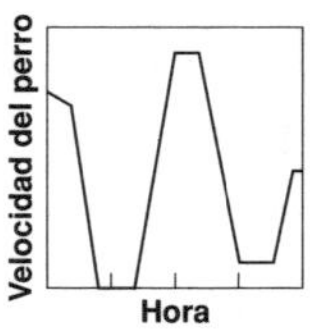

Ejercicio 2 (Perro 3)

4.

Ejercicio 1 (Perro 2)

5. Crea una gráfica que ilustre la temperatura en el interior del automóvil.

Ubicación	Temperatura a la llegada	Temperatura a la salida
Casa	—	74° a las 8:30
Trabajo de verano	77° a las 9:00	128° a las 12:05
Alberca	92° a las 12:15	136° a las 2:30
Biblioteca	95° a las 2:40	77° a las 5:10

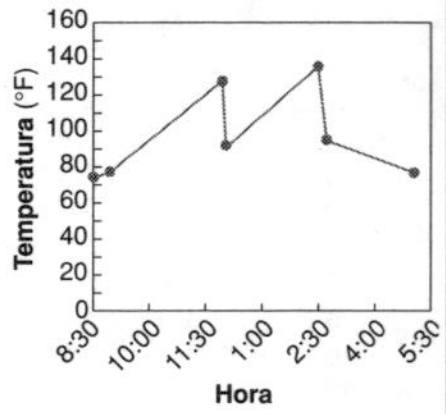

Copyright © by Holt, Rinehart and Winston. All rights reserved.

Copyright © by Holt, Rinehart and Winston. All rights reserved.

LESSON 3-4
Practice
Functions

Complete the table and graph each function.

1. $y = -2x + 5$

x	−2x + 5	y
−2	−2(−2) + 5	9
−1	−2(−1) + 5	7
0	−2(0) + 5	5
1	−2(1) + 5	3
2	−2(2) + 5	1

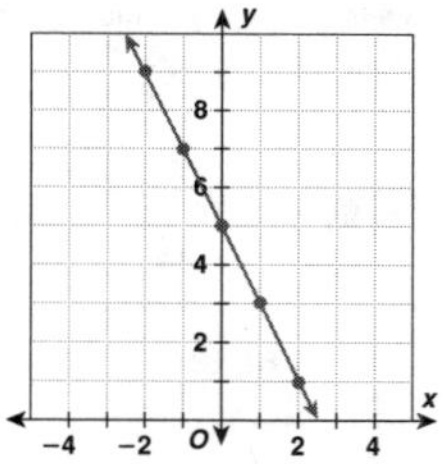

2. $y = x - 2$

x	x − 2	y
−2	−2 − 2	−4
−1	−1 − 2	−3
0	0 − 2	−2
1	1 − 2	−1
2	2 − 2	0

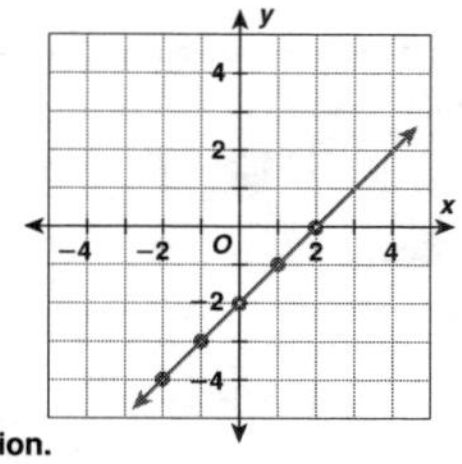

Determine if each relationship represents a function.

3. $y = \frac{1}{3}x - \frac{2}{5}$

yes

4.

x	1	2	1	2
y	6	5	−6	−5

no

5.

x	y
0	0
1	−1
2	−8
3	−27
4	−64

yes

6.

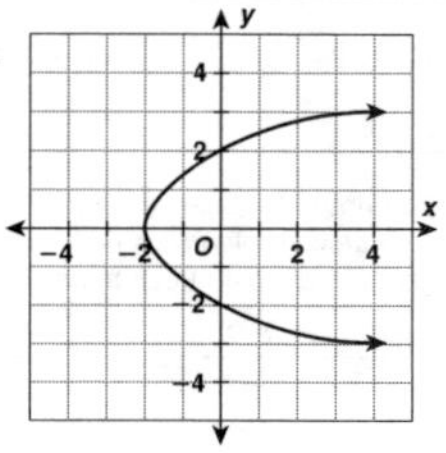

no

Copyright © by Holt, Rinehart and Winston. All rights reserved.

LECCIÓN 3-4
Práctica
Funciones

Completa la tabla y representa gráficamente cada función.

1. $y = -2x + 5$

x	− 2x + 5	y
−2	−2 (−2) + 5	9
−1	−2 (−1) + 5	7
0	−2 (0) + 5	5
1	−2 (1) + 5	3
2	−2 (2) + 5	1

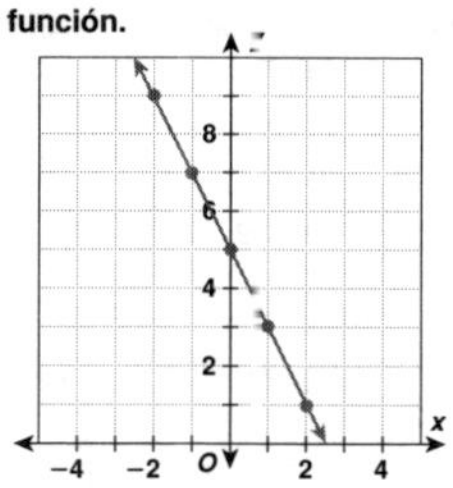

2. $y = x - 2$

x	x − 2	y
−2	−2 − 2	−4
−1	−1 − 2	−3
0	0 − 2	−2
1	1 − 2	−1
2	2 − 2	0

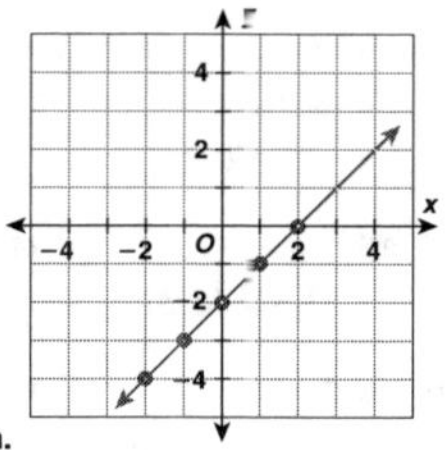

Determina si cada relación representa una función.

3. $y = \frac{1}{3}x - \frac{2}{5}$

sí

4.

x	1	2		2
y	6	5	−6	−5

no

5.

x	y
0	0
1	−1
2	−8
3	−27
4	−64

sí

6.

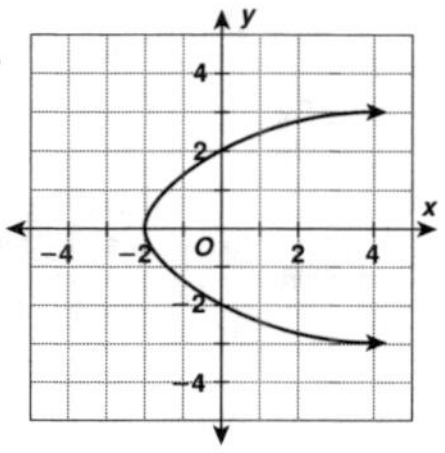

no

Copyright © by Holt, Rinehart and Winston. All rights reserved.

LESSON 3-5
Practice
Equations, Tables, and Graphs

1. The amount of water in a tank being filled is represented by the equation $g = 20m$, where g is the number of gallons in the tank after m minutes. Make a table and sketch a graph of the equation.

m	20m	g
0	20(0)	0
1	20(1)	20
2	20(2)	40
3	20(3)	60
4	20(4)	80

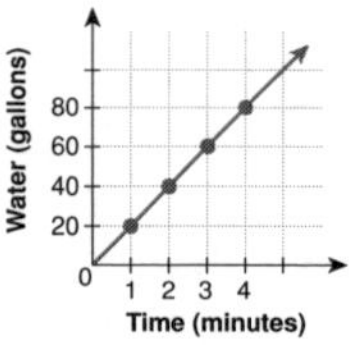

2. Use the table to make a graph and to write an equation.

x	0	2	5	8	12
y	4	6	9	12	16

$y = x + 4$

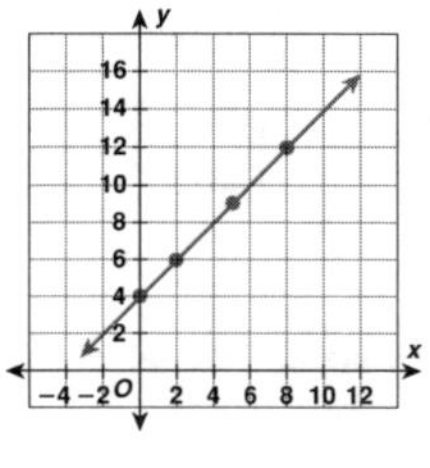

3. Use the graph to make a table and to write an equation.

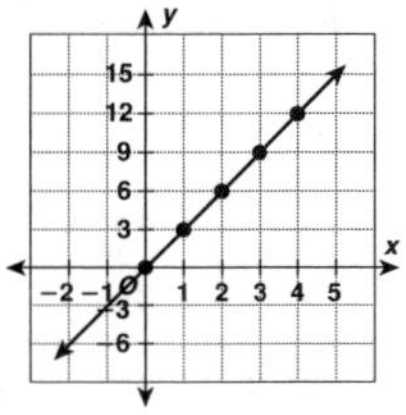

x	0	1	2	3	4
y	0	3	6	9	12

$y = 3x$

Copyright © by Holt, Rinehart and Winston. All rights reserved.

LECCIÓN 3-5
Práctica
Ecuaciones, tablas y gráficas

1. La cantidad de agua en un tanque que se está llenando se representa por medio de la ecuación $g = 20m$, donde g es la cantidad de galones que hay en el tanque después de m minutos. Haz una tabla y dibuja una gráfica de la ecuación.

m	20m	g
0	20(0)	0
1	20(1)	20
2	20(2)	40
3	20(3)	60
4	20(4)	80

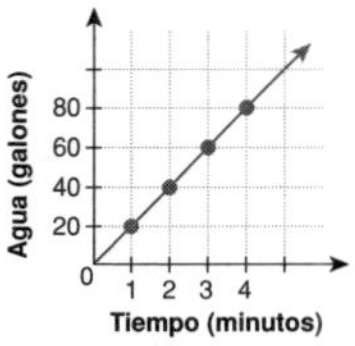

2. Usa la tabla para hacer una gráfica y escribir una ecuación.

x	0	2	5	8	12
y	4	6	9	12	16

$y = x + 4$

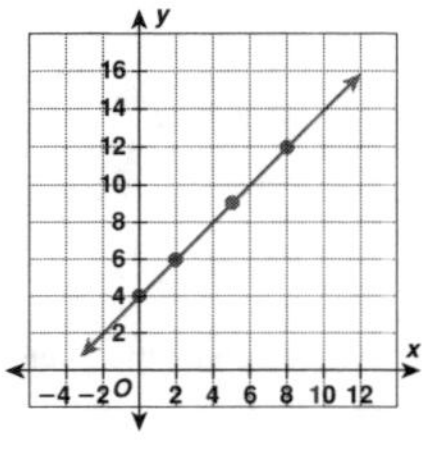

3. Usa la gráfica para hacer una tabla y escribir una ecuación.

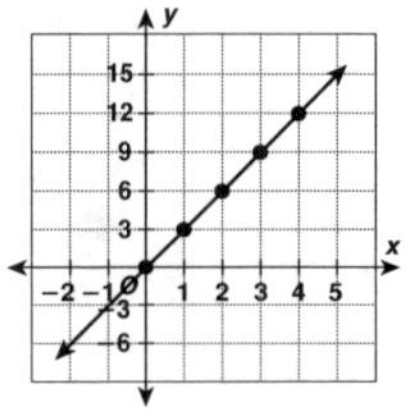

x	0	1	2	3	4
y	0	3	6	9	12

$y = 3x$

Copyright © by Holt, Rinehart and Winston. All rights reserved.

Copyright © by Holt, Rinehart and Winston.
All rights reserved.

LESSON **3-6**

Practice
Arithmetic Sequences

Find the common difference in each arithmetic sequence.

1. 5, 9, 13, 17, ... — 4
2. 3, 10, 17, 24, ... — 7
3. 35, 32, 29, 26, ... — −3
4. 6, 15, 24, 33, ... — 9
5. 92, 87, 82, 77, ... — −5
6. 60, 54, 48, 42, ... — −6
7. 108, 96, 84, 72, ... — −12
8. 3.8, 4, 4.2, 4.4, ... — 0.2
9. 95, 88, 81, 74, ... — −7

Find the next three terms in each arithmetic sequence.

10. 12, 18, 24, 30, ... — 36, 42, 48
11. $1\frac{1}{2}, 2, 2\frac{1}{2}, 3, \ldots$ — $3\frac{1}{2}, 4, 4\frac{1}{2}$
12. −7, −14, −21, −28, ... — −35, −42, −49
13. 0.5, 1, 1.5, 2, ... — 2.5, 3, 3.5
14. −8, −16, −24, −32, ... — −40, −48, −56
15. 72, 63, 54, 45, ... — 36, 27, 18
16. 3.5, 7, 10.5, 14, ... — 17.5, 21, 24.5
17. $\frac{1}{3}, \frac{2}{3}, 1, 1\frac{1}{3}, \ldots$ — $1\frac{2}{3}, 2, 2\frac{1}{3}$
18. 10, 9.5, 9, 8.5, ... — 8, 7.5, 7

Find a function that describes each arithmetic sequence. Use *y* to identify each term in the sequence and *n* to identify each term's position.

19. 6, 12, 18, 24, ... — $y = 6n$
20. −8, −16, −24, −32, ... — $y = -8n$
21. 12, 24, 36, 48, ... — $y = 12n$

22. It costs $12 to rent a mini-car to go around the track, plus $4 per lap. Find a function that describes the sequence. Then find the total cost of driving 5 laps around the track.

$y = 12 + 4n$; $32

Copyright © by Holt, Rinehart and Winston. All rights reserved.

LECCIÓN **3-6**

Práctica
Sucesiones aritméticas

Halla la diferencia común en cada sucesión aritmética.

1. 5, 9, 13, 17, ... — 4
2. 3, 10, 17, 24, ... — 7
3. 35, 32, 29, 26, ... — −3
4. 6, 15, 24, 33, ... — 9
5. 92, 87, 82, 77, ... — −5
6. 60, 54, 48, 42, ... — −6
7. 108, 96, 84, 72, ... — −12
8. 3.8, 4, 4.2, 4.4, ... — 0.2
9. 95, 88, 81, 74, ... — −7

Halla los tres términos siguientes en cada sucesión aritmética.

10. 12, 18, 24, 30, ... — 36, 42, 48
11. $1\frac{1}{2}, 2, 2\frac{1}{2}, 3, \ldots$ — $3\frac{1}{2}, 4, 4\frac{1}{2}$
12. −7, −14, −21, −28, ... — −35, −42, −49
13. 0.5, 1, 1.5, 2, ... — 2.5, 3, 3.5
14. −8, −16, −24, −32, ... — −40, −48, −56
15. 72, 63, 54, 45, ... — 36, 27, 18
16. 3.5, 7, 10.5, 14, ... — 17.5, 21, 24.5
17. $\frac{1}{3}, \frac{2}{3}, 1, 1\frac{1}{3}, \ldots$ — $1\frac{2}{3}, 2, 2\frac{1}{3}$
18. 10, 9.5, 9, 8.5, ... — 8, 7.5, 7

Halla una función que describa cada sucesión aritmética. Usa *y* para identificar cada término en la sucesión y *n* para identificar la posición de cada término.

19. 6, 12, 18, 24, ... — $y = 6n$
20. −8, −16, −24, −32, ... — $y = -8n$
21. 12, 24, 36, 48, ... — $y = 12n$

22. Alquilar un go-kart para dar vueltas en la pista cuesta $12, más $4 por vuelta. Halla una función que describa la sucesión. Luego halla el costo total de dar 5 vueltas en la pista.

$y = 12 + 4n$; $32

Copyright © by Holt, Rinehart and Winston. All rights reserved.

LESSON **4-1**

Practice
Exponents

Write in exponential form.

1. 6 • 6 • 6 • 6 • 6 • 6 — 6^6
2. 7 • 7 • 7 • 7 — 7^4
3. (−8) • (−8) • (−8) • (−8) — $(-8)^4$
4. 5 • 5 • 5 • *b* • *b* • *b* • *b* — $5^3 b^4$

Evaluate.

5. 10^2 — 100
6. $(-6)^2$ — 36
7. 8^2 — 64
8. $(-7)^2$ — 49
9. $(-5)^3$ — −125
10. 12^2 — 144
11. $(-9)^2$ — 81
12. $(-4)^3$ — −64
13. 2^5 — 32
14. 5^4 — 625
15. $(-3)^4$ — 81
16. 6^3 — 216

Evaluate each expression for the given values of the variables.

17. $n^3 - 5$ for $n = 4$ — 59
18. $4x^2 + y^3$ for $x = 5$ and $y = -2$ — 92
19. $m^p + q^2$ for $m = 5$, $p = 2$, and $q = 4$ — 41
20. $a^4 + 2(b - c^2)$ for $a = 2$, $b = 4$, and $c = -1$ — 22

21. Write an expression for five times a number used as a factor three times.

$5x^3$

22. Find the volume of a regular cube if the length of a side is 10 cm. (Hint: $V = l^3$.)

1000 cm^3

Copyright © by Holt, Rinehart and Winston. All rights reserved.

LECCIÓN **4-1**

Práctica
Exponentes

Escribe en forma exponencial.

1. 6 • 6 • 6 • 6 • 6 • 6 — 6^6
2. 7 • 7 • 7 • 7 — 7^4
3. (−8) • (−8) • (−8) • (−8) — $(-8)^4$
4. 5 • 5 • 5 • *b* • *b* • *b* • *b* — $5^3 b^4$

Evalúa.

5. 10^2 — 100
6. $(-6)^2$ — 36
7. 8^2 — 64
8. $(-7)^2$ — 49
9. $(-5)^3$ — −125
10. 12^2 — 144
11. $(-9)^2$ — 81
12. $(-4)^3$ — −64
13. 2^5 — 32
14. 5^4 — 625
15. $(-3)^4$ — 81
16. 6^3 — 216

Evalúa cada expresión para los valores dados de cada variable.

17. $n^3 - 5$ para $n = 4$ — 59
18. $4x^2 + y^3$ para $x = 5$ e $y = -2$ — 92
19. $m^p + q^2$ para $m = 5$, $p = 2$ y $q = 4$ — 41
20. $a^4 + 2(b - c^2)$ para $a = 2$, $b = 4$ y $c = -1$ — 22

21. Escribe una expresión para un número multiplicado por cinco que se usa como factor tres veces.

$5x^3$

22. Halla el volumen de un cubo regular si la longitud de un lado es de 10 cm. (Pista: $V = l^3$.)

1000 cm^3

Copyright © by Holt, Rinehart and Winston. All rights reserved.

Copyright © by Holt, Rinehart and Winston. All rights reserved.

LESSON 4-2
Practice
Look for a Pattern in Integer Exponents

Evaluate the powers of 10.

1. 10^{-3} — $\frac{1}{1,000}$
2. 10^{3} — 1000
3. 10^{-5} — 0.00001
4. 10^{-2} — 0.01
5. 10^{0} — 1
6. 10^{4} — 10,000
7. 10^{1} — 10
8. 10^{5} — 100,000

Evaluate.

9. $(-6)^{-2}$ — $\frac{1}{36}$
10. $(-9)^{-3}$ — $\frac{1}{729}$
11. 2^{-5} — $\frac{1}{32}$
12. $(-3)^{-4}$ — $\frac{1}{81}$
13. $(-12)^{-1}$ — $-\frac{1}{12}$
14. 6^{-3} — $\frac{1}{216}$
15. $10 - (3 + 2)^{0} + 2^{-1}$ — $9\frac{1}{2}$
16. $15 + (-6)^{0} - 3^{-2}$ — $15\frac{8}{9}$
17. $6(8 - 2)^{0} + 4^{-2}$ — $6\frac{1}{16}$
18. $2^{-2} + (-4)^{-1}$ — 0
19. $3(1 - 4)^{-2} + 9^{-1} + 12^{0}$ — $1\frac{4}{9}$
20. $9^{0} + 64(3 + 5)^{-2}$ — 2
21. One milliliter equals 10^{-3} liter. Evaluate 10^{-3}.
    $\frac{1}{1000}$
22. The volume of a cube is 10^{6} cubic feet. Evaluate 10^{6}.
    1,000,000

Copyright © by Holt, Rinehart and Winston.

LECCIÓN 4-2
Práctica
Buscar un patrón en exponentes enteros

Evalúa las potencias de 10.

1. 10^{-3} — $\frac{1}{10,000}$
2. 10^{3} — 1000
3. 10^{-5} — 0.00001
4. 10^{-2} — 0.01
5. 10^{0} — 1
6. 10^{4} — 10,000
7. 10^{1} — 10
8. 10^{5} — 100,000

Evalúa.

9. $(-6)^{-2}$ — $\frac{1}{36}$
10. $(-9)^{-3}$ — $\frac{1}{729}$
11. 2^{-5} — $\frac{1}{32}$
12. $(-3)^{-4}$ — $\frac{1}{81}$
13. $(-12)^{-1}$ — $-\frac{1}{12}$
14. 6^{-3} — $\frac{1}{216}$
15. $10 - (3 + 2)^{0} + 2^{-1}$ — $9\frac{1}{2}$
16. $15 + (-6)^{0} - 3^{-2}$ — $15\frac{8}{9}$
17. $6(8 - 2)^{0} + 4^{-2}$ — $6\frac{1}{16}$
18. $2^{-2} + (-4)^{-1}$ — 0
19. $3(1 - 4)^{-2} + 9^{-1} + 12^{0}$ — $1\frac{4}{9}$
20. $9^{0} + 64(3 + 5)^{-2}$ — 2
21. Un mililitro es igual a 10^{-3} litros. Evalúa 10^{-3}.
    $\frac{1}{1000}$
22. El volumen de un cubo es 10^{6} pies cúbicos. Evalúa 10^{6}.
    1,000,000

Copyright © by Holt, Rinehart and Winston.

LESSON 4-3
Practice
Properties of Exponents

Multiply. Write the product as one power.

1. $10^{5} \cdot 10^{7}$ — 10^{12}
2. $x^{9} \cdot x^{8}$ — x^{17}
3. $14^{7} \cdot 14^{9}$ — 14^{16}
4. $12^{6} \cdot 12^{8}$ — 12^{14}
5. $y^{12} \cdot y^{10}$ — y^{22}
6. $15^{9} \cdot 15^{14}$ — 15^{23}
7. $(-11)^{20} \cdot (-11)^{10}$ — $(-11)^{30}$
8. $(-a)^{6} \cdot (-a)^{7}$ — $(-a)^{13}$

Divide. Write the quotient as one power.

9. $\frac{12^{9}}{12^{2}}$ — 12^{7}
10. $\frac{(-11)^{12}}{(-11)^{8}}$ — $(-11)^{4}$
11. $\frac{x^{10}}{x^{5}}$ — x^{5}
12. $\frac{16^{10}}{16^{2}}$ — 16^{8}
13. $\frac{17^{19}}{17^{2}}$ — 17^{17}
14. $\frac{14^{15}}{14^{13}}$ — 14^{2}
15. $\frac{23^{17}}{23^{9}}$ — 23^{8}
16. $\frac{(-a)^{12}}{(-a)^{7}}$ — $(-a)^{5}$

Simplify.

17. $(6^{2})^{4}$ — 6^{8}
18. $(2^{4})^{-3}$ — 2^{-12}
19. $(3^{5})^{-1}$ — 3^{-5}
20. $(y^{5})^{2}$ — y^{10}
21. $(9^{-2})^{3}$ — 9^{-6}
22. $(10^{0})^{3}$ — 10^{0}
23. $(x^{4})^{-2}$ — x^{-8}
24. $(5^{-2})^{0}$ — 5^{0}

Write the product or quotient as one power.

25. $\frac{w^{12}}{w^{3}}$ — w^{9}
26. $d^{8} \cdot d^{5}$ — d^{13}
27. $(-15)^{5} \cdot (-15)^{10}$ — $(-15)^{15}$
28. Jefferson High School has a student body of 6^{4} students. Each class has approximately 6^{2} students. How many classes does the school have? Write the answer as one power.
    6^{2}
29. Write the expression for a number used as a factor fifteen times being multiplied by a number used as a factor ten times. Then, write the product as one power.
    $x^{15} \cdot x^{10} = x^{25}$

Copyright © by Holt, Rinehart and Winston. All rights reserved.

LECCIÓN 4-3
Práctica
Propiedades de los exponentes

Multiplica. Escribe el producto como una potencia.

1. $10^{5} \cdot 10^{7}$ — 10^{12}
2. $x^{9} \cdot x^{8}$ — x^{17}
3. $14^{7} \cdot 14^{9}$ — 14^{16}
4. $12^{6} \cdot 12^{8}$ — 12^{14}
5. $y^{12} \cdot y^{10}$ — y^{22}
6. $15^{9} \cdot 15^{14}$ — 15^{23}
7. $(-11)^{20} \cdot (-11)^{10}$ — $(-11)^{30}$
8. $(-a)^{6} \cdot (-a)^{7}$ — $(-a)^{13}$

Divide. Escribe el cociente como una potencia.

9. $\frac{12^{9}}{12^{2}}$ — 12^{7}
10. $\frac{(-11)^{12}}{(-11)^{8}}$ — $(-11)^{4}$
11. $\frac{x^{10}}{x^{5}}$ — x^{5}
12. $\frac{16^{10}}{16^{2}}$ — 16^{8}
13. $\frac{17^{19}}{17^{2}}$ — 17^{17}
14. $\frac{14^{15}}{14^{13}}$ — 14^{2}
15. $\frac{23^{17}}{23^{9}}$ — 23^{8}
16. $\frac{(-a)^{12}}{(-a)^{7}}$ — $(-a)^{5}$

Simplifica.

17. $(6^{2})^{4}$ — 6^{8}
18. $(2^{4})^{-3}$ — 2^{-12}
19. $(3^{5})^{-1}$ — 3^{-5}
20. $(y^{5})^{2}$ — y^{10}
21. $(9^{-2})^{3}$ — 9^{-6}
22. $(10^{0})^{3}$ — 10^{0}
23. $(x^{4})^{-2}$ — x^{-8}
24. $(5^{-2})^{0}$ — 5^{0}

Escribe el producto o el cociente como una potencia.

25. $\frac{w^{12}}{w^{3}}$ — w^{9}
26. $d^{8} \cdot d^{5}$ — d^{13}
27. $(-15)^{5} \cdot (-15)^{10}$ — $(-15)^{15}$
28. La escuela secundaria Jefferson tiene un cuerpo estudiantil de 6^{4} estudiantes. En cada clase hay aproximadamente 6^{2} estudiantes. ¿Cuántas clases hay en la escuela? Escribe la respuesta como una potencia.
    6^{2}
29. Escribe la expresión para un número que se usa como factor quince veces multiplicado por un número que se usa como factor diez veces. Luego, escribe el producto como una potencia.
    $x^{15} \cdot x^{10} = x^{25}$

Copyright © by Holt, Rinehart and Winston. All rights reserved.

Copyright © by Holt, Rinehart and Winston.
All rights reserved.

LESSON 4-4

Practice

Scientific Notation

Write each number in standard notation.

1. 2.54×10^2 — 254
2. 6.7×10^{-2} — 0.067
3. 1.14×10^3 — 1140
4. 3.8×10^{-1} — 0.38
5. 7.53×10^{-3} — 0.00753
6. 5.6×10^4 — 56,000
7. 9.1×10^5 — 910,000
8. 6.08×10^{-4} — 0.000608
9. 8.59×10^5 — 859,000
10. 3.331×10^6 — 3,331,000
11. 7.21×10^{-3} — 0.00721
12. 5.88×10^{-4} — 0.000588

Write each number in scientific notation.

13. 75,000,000 — 7.5×10^7
14. 208 — 2.08×10^2
15. 907,100 — 9.071×10^5
16. 56 — 5.6×10^1
17. 0.093 — 9.3×10^{-2}
18. 0.00006 — 6.0×10^{-5}
19. 0.00852 — 8.52×10^{-3}
20. 0.0505 — 5.05×10^{-2}
21. 0.003007 — 3.007×10^{-3}
22. 5226 — 5.226×10^3
23. 0.04 — 4.0×10^{-2}
24. 98,856 — 9.8856×10^4

25. Jupiter is about 778,120,000 kilometers from the Sun. Write this number in scientific notation.

7.7812×10^8

26. The *E. coli* bacterium is about 5×10^{-7} meters wide. A hair is about 1.7×10^{-5} meters wide. Which is wider, the bacterium or the hair?

the hair

Copyright © by Holt, Rinehart and Winston. All rights reserved.

LECCIÓN 4-4

Práctica

Notación científica

Escribe cada número en forma estándar.

1. 2.54×10^2 — 254
2. 6.7×10^{-2} — 0.067
3. 1.14×10^3 — 1140
4. 3.8×10^{-1} — 0.38
5. 7.53×10^{-3} — 0.00753
6. 5.6×10^4 — 56,000
7. 9.1×10^5 — 910,000
8. 6.08×10^{-4} — 0.000608
9. 8.59×10^5 — 859,000
10. 3.331×10^6 — 3,331,000
11. 7.21×10^{-3} — 0.00721
12. 5.88×10^{-4} — 0.000588

Escribe cada número en notación científica.

13. 75,000,000 — 7.5×10^7
14. 208 — 2.08×10^2
15. 907,100 — 9.071×10^5
16. 56 — 5.6×10^1
17. 0.093 — 9.3×10^{-2}
18. 0.00006 — 6.0×10^{-5}
19. 0.00852 — 8.52×10^{-3}
20. 0.0505 — 5.05×10^{-2}
21. 0.003007 — 3.007×10^{-3}
22. 5226 — 5.226×10^3
23. 0.04 — 4.0×10^{-2}
24. 98,856 — 9.8856×10^4

25. Júpiter está a aproximadamente 778,120,000 kilómetros del Sol. Escribe este número en notación científica.

7.7812×10^8

26. La bacteria *E. coli* mide aproximadamente 5×10^{-7} metros de ancho. Un pelo mide aproximadamente 1.7×10^{-5} metros de ancho. ¿Qué es más ancho: la bacteria o el pelo?

el pelo

Copyright © by Holt, Rinehart and Winston. All rights reserved.

LESSON 4-5

Practice

Squares and Square Roots

Find the two square roots of each number.

1. 36 — 6, −6
2. 81 — 9, −9
3. 49 — 7, −7
4. 100 — 10, −10
5. 64 — 8, −8
6. 121 — 11, −11
7. 25 — 5, −5
8. 144 — 12, −12

Evaluate each expression.

9. $\sqrt{32 + 17}$ — 7
10. $\sqrt{100 - 19}$ — 9
11. $\sqrt{64 + 36}$ — 10
12. $\sqrt{73 - 48}$ — 5
13. $2\sqrt{64} + 10$ — 26
14. $36 - \sqrt{36}$ — 30
15. $\sqrt{100} - \sqrt{25}$ — 5
16. $\sqrt{121} + 16$ — 27
17. $\sqrt{\frac{25}{4}} + \frac{1}{2}$ — 3
18. $\sqrt{\frac{100}{25}}$ — 2
19. $\sqrt{\frac{196}{49}}$ — 2
20. $3(\sqrt{144} - 6)$ — 18

The Pyramids of Egypt are often called the first wonder of the world. This group of pyramids consists of Menkaura, Khufu, and Khafra. The largest of these is Khufu, sometimes called Cheops. During this time in history, each monarch had his own pyramid built to bury his mummified body. Cheops was a king of Egypt in the early 26th century B.C. His pyramid's original height is estimated to have been 482 ft. It is now approximately 450 ft. The estimated completion date of this structure was 2660 B.C.

21. If the area of the base of Cheops' pyramid is 570,025 ft^2, what is the length of one of the sides of the ancient structure? (Hint: $s = \sqrt{A}$)

755 ft

22. If a replica of the pyramid were built with a base area of 625 in^2, what would be the length of each side? (Hint: $s = \sqrt{A}$)

25 in.

Copyright © by Holt, Rinehart and Winston. All rights reserved.

LECCIÓN 4-5

Práctica

Cuadrados y raíces cuadradas

Halla las dos raíces cuadradas de cada número.

1. 36 — 6, −6
2. 81 — 9, −9
3. 49 — 7, −7
4. 100 — 10, −10
5. 64 — 8, −8
6. 121 — 11, −11
7. 25 — 5, −5
8. 144 — 12, −12

Evalúa cada expresión.

9. $\sqrt{32 + 17}$ — 7
10. $\sqrt{100 - 19}$ — 9
11. $\sqrt{64 + 36}$ — 10
12. $\sqrt{73 - 48}$ — 5
13. $2\sqrt{64} + 10$ — 26
14. $36 - \sqrt{36}$ — 30
15. $\sqrt{100} - \sqrt{25}$ — 5
16. $\sqrt{121} + 16$ — 27
17. $\sqrt{\frac{25}{4}} + \frac{1}{2}$ — 3
18. $\sqrt{\frac{100}{25}}$ — 2
19. $\sqrt{\frac{196}{49}}$ — 2
20. $3(\sqrt{144} - 6)$ — 18

A menudo, a las pirámides de Egipto se las llama la primera maravilla del mundo. Este grupo de pirámides está compuesto por las de Menkura, Khufu y Khafra. La más grande de las tres es la de Khufu, a veces llamada la pirámide de Keops. Durante ese período de la historia, a cada monarca se le construía su propia pirámide para enterrar su cuerpo momificado. Keops fue rey de Egipto a comienzos del siglo XXVI a.C. Se calcula que la altura original de la pirámide de Keops era de 482 pies. En la actualidad, su altura es de aproximadamente 450 pies. La fecha estimada de la finalización de esta estructura es 2660 a.C.

21. Si el área de la base de la pirámide de Keops es de 570,025 $pies^2$, ¿cuál es la longitud de uno de los lados de la antigua estructura? (Pista: $l = \sqrt{A}$)

755 pies

22. Si se construyera una réplica de la pirámide con un área de base de 625 $pulg^2$, ¿cuál sería la longitud de cada lado? (Pista: $l = \sqrt{A}$)

25 pulg

Copyright © by Holt, Rinehart and Winston. All rights reserved.

Copyright © by Holt, Rinehart and Winston. All rights reserved.

LESSON 4-6
Practice
Estimating Square Roots

Each square root is between two integers. Name the integers. Explain your answer.

1. $\sqrt{6}$
   2 and 3; 6 is between 4 and 9
2. $\sqrt{20}$
   4 and 5; 20 is between 16 and 25
3. $\sqrt{28}$
   5 and 6; 28 is between 25 and 36
4. $\sqrt{44}$
   6 and 7; 44 is between 36 and 49
5. $\sqrt{31}$
   5 and 6; 31 is between 25 and 36
6. $\sqrt{52}$
   7 and 8; 52 is between 49 and 64

Use a calculator to find each value. Round to the nearest tenth.

7. $\sqrt{14}$ 3.7
8. $\sqrt{42}$ 6.5
9. $\sqrt{21}$ 4.6
10. $\sqrt{47}$ 6.9
11. $\sqrt{58}$ 7.6
12. $\sqrt{60}$ 7.7
13. $\sqrt{35}$ 5.9
14. $\sqrt{75}$ 8.7

Police use the formula $r = 2\sqrt{5L}$ to approximate the rate of speed in miles per hours of a vehicle from its skid marks, where L is the length of the skid marks in feet.

15. About how fast is a car going that leaves skid marks of 80 ft?
    40 mi/h
16. About how fast is a car going that leaves skid marks of 245 ft?
    70 mi/h
17. If the formula for finding the length of the skid marks is $L = \frac{r^2}{20}$, what would be the length of the skid marks from a vehicle traveling 80 mi/h?
    320 ft

Copyright © by Holt, Rinehart and Winston. All rights reserved.

LECCIÓN 4-6
Práctica
Cómo estimar raíces cuadradas

Cada raíz cuadrada está entre dos números enteros. Identifica los enteros. Explica tu respuesta.

1. $\sqrt{6}$
   2 y 3; 6 está entre 4 y 9
2. $\sqrt{20}$
   4 y 5; 20 está entre 16 y 25
3. $\sqrt{28}$
   5 y 6; 28 está entre 25 y 36
4. $\sqrt{44}$
   6 y 7; 44 está entre 36 y 49
5. $\sqrt{31}$
   5 y 6; 31 está entre 25 y 36
6. $\sqrt{52}$
   7 y 8; 52 está entre 49 y 64

Usa una calculadora para hallar cada valor. Redondea a la décima más cercana.

7. $\sqrt{14}$ 3.7
8. $\sqrt{42}$ 6.5
9. $\sqrt{21}$ 4.6
10. $\sqrt{47}$ 6.9
11. $\sqrt{58}$ 7.6
12. $\sqrt{60}$ 7.7
13. $\sqrt{35}$ 5.9
14. $\sqrt{75}$ 8.7

La policía usa la fórmula $t = 2\sqrt{5L}$ para calcular la tasa de velocidad de un vehículo en millas por hora a partir de las huellas de frenado, donde L es la longitud de las huellas de frenado en pies.

15. ¿Aproximadamente a qué velocidad circula un vehículo que deja huellas de frenado de 80 pies?
    40 mi/h
16. ¿Aproximadamente a qué velocidad circula un vehículo que deja huellas de frenado de 245 pies?
    70 mi/h
17. Si la fórmula para hallar la longitud de las huellas de frenado es $L = \frac{r^2}{20}$, ¿cuál sería la longitud de las huellas de frenado que dejaría un vehículo que circula a 80 mi/h?
    320 pies

Copyright © by Holt, Rinehart and Winston. All rights reserved.

LESSON 4-7
Practice
The Real Numbers

Write all names that apply to each number.

1. $-\frac{7}{8}$
   rational; real
2. $\sqrt{0.15}$
   irrational; real
3. $\sqrt{\frac{18}{2}}$
   whole; integer; rational; real
4. $\sqrt{45}$
   irrational; real
5. -25
   integer; rational; real
6. -6.75
   rational; real

State if the number is rational, irrational, or not a real number.

7. $\sqrt{14}$ irrational
8. $\sqrt{-16}$ not real
9. $\frac{6.2}{0}$ not real
10. $\sqrt{49}$ rational
11. $\frac{7}{20}$ rational
12. $-\sqrt{81}$ rational
13. $\sqrt{\frac{7}{9}}$ irrational
14. -1.3 rational

Find a real number between each pair of numbers.

15. $7\frac{3}{5}$ and $7\frac{4}{5}$
    sample answer: $7\frac{7}{10}$
16. 6.45 and $\frac{13}{2}$
    sample answer: 6.48
17. $\frac{7}{8}$ and $\frac{9}{10}$
    sample answer: $\frac{22}{25}$
18. Give an example of a rational number between $-\sqrt{4}$ and $\sqrt{4}$
    sample answer: 0
19. Give an example of an irrational number less than 0.
    sample answer: $-\frac{\sqrt{22}}{7}$
20. Give an example of a number that is not real.
    sample answer: $\frac{2}{0}$

Copyright © by Holt, Rinehart and Winston. All rights reserved.

LECCIÓN 4-7
Práctica
Los números reales

Escribe todos los nombres que correspondan a cada número.

1. $-\frac{7}{8}$
   racional; real
2. $\sqrt{0.15}$
   irracional; real
3. $\sqrt{\frac{18}{2}}$
   cabal; entero; racional; real
4. $\sqrt{45}$
   irracional; real
5. -25
   entero; racional; real
6. -6.75
   racional; real

Indica si cada número es racional, irracional o no es un número real.

7. $\sqrt{14}$ irracional
8. $\sqrt{-16}$ no es real
9. $\frac{6.2}{0}$ no es real
10. $\sqrt{49}$ racional
11. $\frac{7}{20}$ racional
12. $-\sqrt{81}$ racional
13. $\sqrt{\frac{7}{9}}$ irracional
14. -1.3 racional

Halla un número real entre cada par de números.

15. $7\frac{3}{5}$ y $7\frac{4}{5}$
    ejemplo de respuesta: $7\frac{7}{10}$
16. 6.45 y $\frac{13}{2}$
    ejemplo de respuesta: 6.48
17. $\frac{7}{8}$ y $\frac{9}{10}$
    ejemplo de respuesta: $\frac{22}{25}$
18. Da un ejemplo de un número racional entre $-\sqrt{4}$ y $\sqrt{4}$
    ejemplo de respuesta: 0
19. Da un ejemplo de un número irracional menor que 0.
    ejemplo de respuesta: $-\frac{\sqrt{22}}{7}$
20. Da un ejemplo de un número que no sea real.
    ejemplo de respuesta $\frac{2}{0}$

Copyright © by Holt, Rinehart and Winston. All rights reserved.

Copyright © by Holt, Rinehart and Winston. All rights reserved.

LESSON **4-8**

Practice
The Pythagorean Theorem

Find the length of the hypotenuse to the nearest tenth.

1. 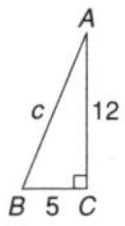

13

2. 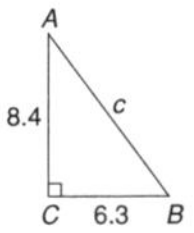

10.5

3. 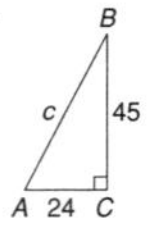

51

Solve for the unknown side in each right triangle to the nearest tenth.

4. 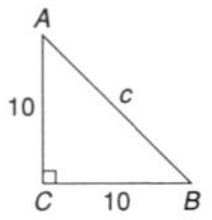

14.1

5. 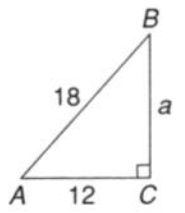

13.4

6. 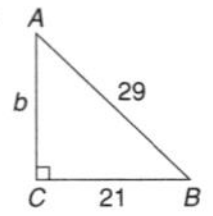

20

7. 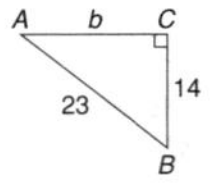

18.2

8. 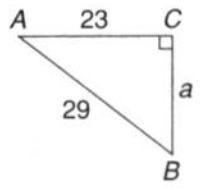

17.7

9. 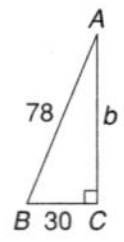

72

10. A glider flies 8 miles south from the airport and then 15 miles east. Then it flies in a straight line back to the airport. What was the distance of the glider's last leg back to the airport?

17 mi

LECCIÓN **4-8**

Práctica
El Teorema de Pitágoras

Halla la longitud de la hipotenusa a la décima más cercana.

1. 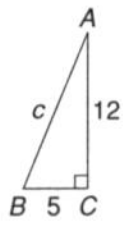

13

2. 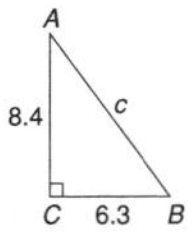

10.5

3. 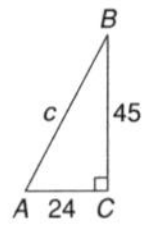

51

Resuelve el lado desconocido del triángulo a la décima más cercana.

4. 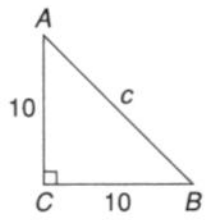

14.1

5. 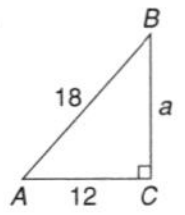

13.4

6. 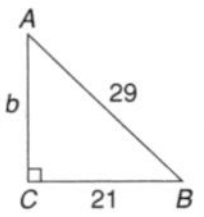

20

7. 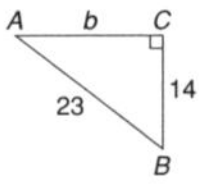

18.2

8. 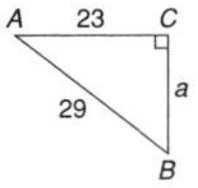

17.7

9. 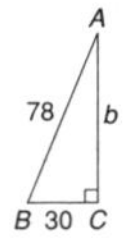

72

10. Un planeador vuela 8 millas hacia el sur desde el aeropuerto y luego 15 millas hacia el este. Después vuela en línea recta de regreso al aeropuerto. ¿Cuál fue la distancia que recorrió el planeador en la última etapa, el regreso al aeropuerto?

17 mi

LESSON **5-1**

Practice
Ratios and Proportions

Find two ratios that are equivalent to each given ratio. Sample answers given.

1. $\frac{9}{12}$ — $\frac{3}{4}, \frac{18}{24}$
2. $\frac{4}{20}$ — $\frac{1}{5}, \frac{6}{30}$
3. $\frac{15}{25}$ — $\frac{3}{5}, \frac{6}{10}$
4. $\frac{7}{12}$ — $\frac{14}{24}, \frac{21}{36}$
5. $\frac{14}{7}$ — $\frac{2}{1}, \frac{8}{4}$
6. $\frac{11}{22}$ — $\frac{1}{2}, \frac{5}{10}$
7. $\frac{10}{3}$ — $\frac{20}{6}, \frac{30}{9}$
8. $\frac{18}{28}$ — $\frac{9}{14}, \frac{27}{42}$
9. $\frac{12}{27}$ — $\frac{4}{9}, \frac{8}{18}$

Simplify to tell whether the ratios form a proportion.

10. $\frac{13}{39}$ and $\frac{16}{48}$ — yes, $\frac{1}{3} = \frac{1}{3}$
11. $\frac{21}{49}$ and $\frac{28}{56}$ — no, $\frac{3}{7} \neq \frac{1}{2}$
12. $\frac{12}{28}$ and $\frac{18}{42}$ — yes, $\frac{3}{7} = \frac{3}{7}$
13. $\frac{18}{27}$ and $\frac{10}{15}$ — yes, $\frac{2}{3} = \frac{2}{3}$
14. $\frac{24}{27}$ and $\frac{27}{30}$ — no, $\frac{8}{9} \neq \frac{9}{10}$
15. $\frac{14}{10}$ and $\frac{35}{25}$ — yes, $\frac{7}{5} = \frac{7}{5}$
16. $\frac{10}{32}$ and $\frac{25}{80}$ — yes, $\frac{5}{16} = \frac{5}{16}$
17. $\frac{16}{48}$ and $\frac{15}{45}$ — yes, $\frac{1}{3} = \frac{1}{3}$

18. Mrs. Walters wanted one daffodil plant for every 2 tulip plants in her garden. If she planted 20 daffodil bulbs, how many tulip bulbs did she plant?

40 tulip bulbs

19. In a survey, 9 out of 10 doctors recommended a certain medicine. If 80 doctors were surveyed, how many doctors recommended the medicine?

72 doctors

20. A molecule of sodium carbonate contains 2 atoms of sodium to every 3 atoms of oxygen. Could a compound containing 12 atoms of sodium and 15 atoms of oxygen be sodium carbonate? Explain.

No, the ratio $\frac{12}{15}$ is not equivalent to $\frac{2}{3}$.

LECCIÓN **5-1**

Práctica
Razones y proporciones

Halla dos razones que sean equivalentes a cada razón dada. Se dan ejemplos de respuestas.

1. $\frac{9}{12}$ — $\frac{3}{4}, \frac{18}{24}$
2. $\frac{4}{20}$ — $\frac{1}{5}, \frac{6}{30}$
3. $\frac{15}{25}$ — $\frac{3}{5}, \frac{6}{10}$
4. $\frac{7}{12}$ — $\frac{14}{24}, \frac{21}{36}$
5. $\frac{14}{7}$ — $\frac{2}{1}, \frac{8}{4}$
6. $\frac{11}{22}$ — $\frac{1}{2}, \frac{5}{10}$
7. $\frac{10}{3}$ — $\frac{20}{6}, \frac{30}{9}$
8. $\frac{18}{28}$ — $\frac{9}{14}, \frac{27}{42}$
9. $\frac{12}{27}$ — $\frac{4}{9}, \frac{8}{18}$

Simplifica para indicar si las razones forman una proporción.

10. $\frac{13}{39}$ y $\frac{16}{48}$ — sí, $\frac{1}{3} = \frac{1}{3}$
11. $\frac{21}{49}$ y $\frac{28}{56}$ — no, $\frac{3}{7} \neq \frac{1}{2}$
12. $\frac{12}{28}$ y $\frac{18}{42}$ — sí, $\frac{3}{7} = \frac{3}{7}$
13. $\frac{18}{27}$ y $\frac{10}{15}$ — sí, $\frac{2}{3} = \frac{2}{3}$
14. $\frac{24}{27}$ y $\frac{27}{30}$ — no, $\frac{8}{9} \neq \frac{9}{10}$
15. $\frac{14}{10}$ y $\frac{35}{25}$ — sí, $\frac{7}{5} = \frac{7}{5}$
16. $\frac{10}{32}$ y $\frac{25}{80}$ — sí, $\frac{5}{16} = \frac{5}{16}$
17. $\frac{16}{48}$ y $\frac{15}{45}$ — sí, $\frac{1}{3} = \frac{1}{3}$

18. La señora Walters quería que en su jardín hubiera un narciso por cada dos tulipanes. Si plantó 20 bulbos de narcisos, ¿cuántos bulbos de tulipán plantó?

40 bulbos de tulipán

19. En una encuesta, 9 de cada 10 médicos recomendaron un medicamento. Si se encuestó a 80 médicos, ¿cuántos médicos recomendaron el medicamento?

72 médicos

20. Una molécula de carbonato de sodio contiene 2 átomos de sodio cada 3 átomos de oxígeno. ¿Es posible que un compuesto que contiene 12 átomos de sodio y 15 átomos de oxígeno sea carbonato de sodio? Explica.

No, la razón $\frac{12}{15}$ no equivale a $\frac{2}{3}$.

LESSON **5-2**

Practice
Ratios, Rates, and Unit Rates

1. Copper weighing 4480 kilograms has a volume of 0.5 cubic meters. What is the density of copper?

   $\frac{8960 \text{ kg}}{\text{m}^3}$

2. Yoshi's yogurt contains 15 calories per ounce. How many calories are in an 8-ounce container of Yoshi's yogurt?

   120 calories

3. Emily earns $7.50 per hour. How much does she earn in 3 hours?

   $22.50

Estimate the unit rate.

4. 43 apples in 5 bags

   about 9 apples per bag

5. $71.00 for 8 hours

   about $9 per hour

6. 146 students in 6 classes

   about 25 students per class

7. $52.00 for 5 hours

   about $10 per hour

8. 7 miles in 64 minutes

   about 9 minutes per mile

9. $3.55 for 4 pounds

   about $0.90 per pound

Determine the better buy.

10. 8.2 oz of toothpaste for $2.99 or 6.4 oz of toothpaste for $2.49

    8.2 oz for $2.99

11. a 3 lb bag of apples for $2.99 or a 5 lb bag of apples for $4.99

    3 lb bag for $2.99

12. 16 oz bottle of soda for $1.25 or 20 oz bottle of soda for $1.55

    20 oz bottle for $1.55

13. Mavis rides the bus every day. She bought a bus pass good for the month of October for $38.75. How much was Mavis charged per day for the bus pass?

    $1.25

Copyright © by Holt, Rinehart and Winston. All rights reserved.

LECCIÓN **5-2**

Práctica
Razones, tasas y tasas unitarias

1. Un trozo de cobre que pesa 4480 kilogramos tiene un volumen de 0.5 metros cúbicos. ¿Cuál es la densidad del cobre?

   $\frac{8960 \text{ Kg}^3}{\text{m}^3}$

2. El yogur Yoshi tiene 15 calorías por onza. ¿Cuántas calorías tiene un envase de 8 onzas de yogur?

   120 calorías

3. Emily gana $7.50 por hora. ¿Cuánto gana en 3 horas?

   $22.50

Estima la tasa unitaria.

4. 43 manzanas en 5 bolsas

   aproximadamente 9 manzanas por bolsa

5. $71.00 por 8 horas

   aproximadamente $9 por hora

6. 146 estudiantes en 6 clases

   aproximadamente 25 estudiantes por clase

7. $52.00 por 5 horas

   aproximadamente $10 por hora

8. 7 millas en 64 minutos

   aproximadamente 9 minutos por milla

9. $3.55 por 4 libras

   aproximadamente $0 90 por libra

Determina cuál es la mejor opción de compra.

10. Un dentífrico de 8.2 oz a $2.99 o un dentífrico de 6.4 oz a $2.49

    8.2 oz a $2.99

11. Una bolsa de manzanas de 3 lb a $2.99 o una bolsa de manzanas de 5 lb a $4.99

    una bolsa de 3 lb a $2.99

12. Una botella de refresco de 16 oz a $1.25 o una botella de refresco de 20 oz a $1.55

    una botella de 20 oz a $1.55

13. Mavis toma el autobús todos los días. Compró el abono de autobús para el mes de octubre a $38.75. ¿Cuánto le cobraron por día por el abono de autobús?

    $1.25

Copyright © by Holt, Rinehart and Winston. All rights reserved.

LESSON **5-3**

Practice
Dimensional Analysis

Find the appropriate factor for each conversion.

1. grams to kilograms

   $\frac{1 \text{ kg}}{1000 \text{ g}}$

2. quarts to gallons

   $\frac{1 \text{ gal}}{4 \text{ qt}}$

3. minutes to seconds

   $\frac{60 \text{ sec}}{1 \text{ min}}$

4. David takes 300 milligrams of medicine every day. How many grams is this?

   0.3 g

5. Jody runs the 500-yard dash for his school's track team. How many feet does he run in each 500-yard dash?

   1500 ft

6. Sean drinks six 12-ounce cans of soda a week. How many pints of soda does he drink in a week?

   4.5 pt

7. A recipe for punch requires diluting the punch concentrate with 7 quarts of water. How many gallons of water are required to dilute the concentrate according to the directions?

   1.75 gal

8. Jesse's dog Angel weighs $18\frac{1}{2}$ pounds. How many ounces does Angel weigh?

   296 oz

9. A roll of tape contains 32.9 meters of tape. How many millimeters of tape does the roll contain?

   32,900 mm

10. There are two types of lifts in the sport of weightlifting, the *snatch* and the *clean and jerk*. Winners are determined by the combined weights of the two type of lifts. In the 2002 Collegiate Weightlifting Competition, Timothy Leancu from the U.S. Naval Academy competed in the 94-kilogram weight class. He lifted 100 kg in the *snatch* and 132.5 kg in the *clean and jerk*. What was the combined weight of his lifts in grams?

    232,500 g

Copyright © by Holt, Rinehart and Winston. All rights reserved.

LECCIÓN **5-3**

Práctica
Análisis dimensional

Halla el factor apropiado para cada conversión.

1. gramos en kilogramos

   $\frac{1 \text{ kg}}{1000 \text{ g}}$

2. cuartos en galones

   $\frac{1 \text{ gal}}{4 \text{ ct}}$

3. minutos en segundos

   $\frac{60 \text{ s}}{1 \text{ min}}$

4. David toma 300 miligramos de medicamento por día. ¿Cuántos gramos son?

   0.3 g

5. Jody corre las 500 yardas planas para el equipo de atletismo de la escuela. ¿Cuántos pies corre en cada carrera?

   1500 pies

6. Sean bebe seis latas de refresco de 12 onzas por semana. ¿Cuántas pintas de refresco bebe en una semana?

   4.5 pt

7. Una receta de ponche indica que hay que diluir el concentrado de ponche en 7 cuartos de agua. ¿Cuántos galones de agua se necesitan para diluir el concentrado de acuerdo con la receta?

   1.75 gal

8. Ángel, el perro de Jesse, pesa $18\frac{1}{2}$ libras. ¿Cuántas onzas pesa?

   296 oz

9. Un rollo de cinta tiene 32.9 metros de cinta. ¿Cuántos milímetros tiene?

   32,900 mm

10. En el levantamiento de pesas, hay dos tipos de levantamientos: *arranque* y *envión*. Se determina el ganador por el peso combinado de los dos tipos de levantamiento. En la Competencia Universitaria de Levantamiento de Pesas de 2002, Timothy Leancu, de la Academia Naval de EE.UU., compitió en la clase de 94 kilogramos. Levantó 100 kg en el *arranque* y 132.5 kg en el *envión*. ¿Cuál fue el peso combinado de los levantamientos en gramos?

    232,500 g

Copyright © by Holt, Rinehart and Winston. All rights reserved.

Copyright © by Holt, Rinehart and Winston. All rights reserved.

LESSON 5-4

Practice
Solving Proportions

Tell whether the ratios are proportional.

1. $\frac{3}{4} \stackrel{?}{=} \frac{9}{12}$ yes
2. $\frac{9}{24} \stackrel{?}{=} \frac{18}{48}$ yes
3. $\frac{16}{24} \stackrel{?}{=} \frac{10}{18}$ no
4. $\frac{13}{25} \stackrel{?}{=} \frac{26}{50}$ yes
5. $\frac{10}{32} \stackrel{?}{=} \frac{16}{38}$ no
6. $\frac{20}{36} \stackrel{?}{=} \frac{50}{90}$ yes
7. $\frac{20}{28} \stackrel{?}{=} \frac{28}{36}$ no
8. $\frac{14}{42} \stackrel{?}{=} \frac{16}{36}$ no

Solve each proportion.

9. $\frac{\$d}{3 \text{ CDs}} = \frac{\$64.75}{5 \text{ CDs}}$ $38.85
10. $\frac{c \text{ chairs}}{7 \text{ rows}} = \frac{252 \text{ chairs}}{9 \text{ rows}}$ 196 chairs
11. $\frac{m \text{ miles}}{5 \text{ hours}} = \frac{135 \text{ miles}}{3 \text{ hours}}$ 225 mi
12. $\frac{\$d}{4 \text{ subs}} = \frac{\$45}{10 \text{ subs}}$ $18

Solve each proportional situation using equivalent fractions.

13. $\frac{c}{15} = \frac{4}{10}$ $c = 6$
14. $\frac{a}{6} = \frac{8}{12}$ $a = 4$
15. $\frac{b}{20} = \frac{15}{12}$ $b = 25$
16. $\frac{w}{6} = \frac{15}{10}$ $w = 9$

17. Janessa bought 4 stamps for $1.48. At this rate, how much would 10 stamps cost?

$3.70

18. A karate team had 6 girls and 9 boys. Then 2 more girls and 3 more boys joined the team. Did the ratio of girls to boys stay the same? Explain.

Yes; $\frac{6}{9} = \frac{8}{12}$

19. A 30 kg weight is positioned 2 m from a fulcrum. At what distance from the fulcrum must a 40 kg weight be positioned to keep the scale balanced?

1.5 m

Copyright © by Holt, Rinehart and Winston. All rights reserved.

LECCIÓN 5-4

Práctica
Cómo resolver proporciones

Indica si las razones son proporcionales.

1. $\frac{3}{4} \stackrel{?}{=} \frac{9}{12}$ sí
2. $\frac{9}{24} \stackrel{?}{=} \frac{18}{48}$ sí
3. $\frac{16}{24} \stackrel{?}{=} \frac{10}{18}$ no
4. $\frac{13}{25} \stackrel{?}{=} \frac{26}{50}$ sí
5. $\frac{10}{32} \stackrel{?}{=} \frac{16}{38}$ no
6. $\frac{20}{36} \stackrel{?}{=} \frac{50}{90}$ sí
7. $\frac{20}{28} \stackrel{?}{=} \frac{28}{36}$ no
8. $\frac{14}{42} \stackrel{?}{=} \frac{16}{36}$ no

Resuelve cada proporción.

9. $\frac{\$d}{3 \text{ CD}} = \frac{\$64.75}{5 \text{ CD}}$ $38.85
10. $\frac{c \text{ sillas}}{7 \text{ filas}} = \frac{252 \text{ sillas}}{9 \text{ filas}}$ 196 sillas
11. $\frac{m \text{ millas}}{5 \text{ horas}} = \frac{135 \text{ millas}}{3 \text{ horas}}$ 225 mi
12. $\frac{\$d}{4 \text{ cuotas}} = \frac{\$45}{10 \text{ cuotas}}$ $18

Resuelve cada proporción usando fracciones equivalentes.

13. $\frac{c}{15} = \frac{4}{10}$ $c = 6$
14. $\frac{a}{6} = \frac{8}{12}$ $a = 4$
15. $\frac{b}{20} = \frac{15}{12}$ $b = 25$
16. $\frac{w}{6} = \frac{15}{10}$ $w = 9$

17. Janessa compró 4 estampillas por $1.48. A esta tasa, ¿cuánto costarían 10 estampillas?

$3.70

18. En un equipo de karate había 6 chicas y 9 chicos. Luego se unieron al equipo 2 chicas y 3 chicos más. ¿Se mantuvo la razón de las chicas con respecto a los chicos? Explica.

Sí; $\frac{6}{9} = \frac{8}{12}$

19. Se pone una pesa de 30 kg a 2 m de un fulcro. ¿A qué distancia del fulcro se debe poner una pesa de 40 kg para mantener el equilibrio de la balanza?

1.5 m

Copyright © by Holt, Rinehart and Winston. All rights reserved.

LESSON 5-5

Practice
Similar Figures

1. Are any of these triangles similar?

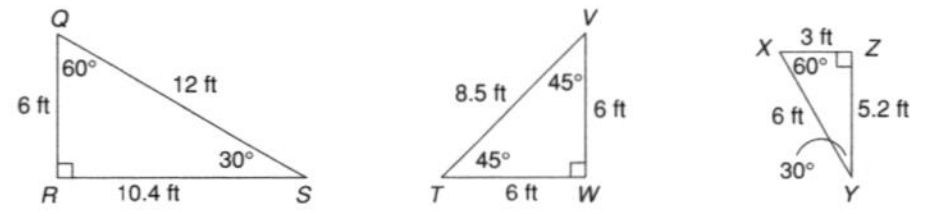

$\triangle QRS \sim \triangle XZY$

2. A photo is 12 in. wide by 18 in. tall. If the width is scaled down to 9 inches, how tall should the similar photo be? 13.5 in.

3. An isosceles triangle has a base of 20 cm and legs measuring 36 cm. How long are the legs of a similar triangle with base measuring 50 cm? 90 cm

4. A picture of a school's mascot is 18 in. wide and 24 in. long. It is enlarged proportionally to banner size. If the width is enlarged to 63 in., what is the length of the banner? 84 in.

5. Carol has a 24 cm × 36 cm photo that she reduces to $\frac{3}{4}$ of its size. What are the dimensions of the new photo? 18 cm × 27 cm

6. Erik is drawing a picture of his school's basketball court. The actual basketball court is 84 ft long and 50 ft wide. If Erik draws the court with a length of 21 in., what will be the width? 12.5 in.

7. IMAX theaters have the world's largest screens. There are numerous IMAX theaters around the world. The Henry Ford Museum in Dearborn, Michigan hosts an IMAX theater with a 60 ft × 84 ft screen. If a classroom projection screen were changed to be in direct proportion with the IMAX screen at the Henry Ford Museum, the dimensions would be 5 ft × ___ ft. 7

Copyright © by Holt, Rinehart and Winston. All rights reserved.

LECCIÓN 5-5

Práctica
Figuras semejantes

1. ¿Algunos de estos triángulos son semejantes?

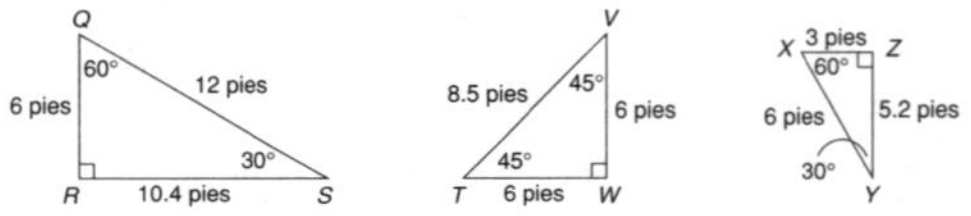

$\triangle QRS \sim \triangle XZY$

2. Una fotografía mide 12 pulg de ancho por 18 pulg de alto. Si el ancho se redujera a 9 pulgadas, ¿qué altura debería tener la fotografía semejante? 13.5 pulg

3. La base de un triángulo isósceles mide 20 cm y los catetos miden 36 cm. ¿Cuánto medirán los catetos de un triángulo semejante cuya base mide 50 cm? 90 cm

4. Una imagen de la mascota de la escuela mide 18 pulg de ancho y 24 pulg de largo. Se la agranda proporcionalmente a un tamaño de pancarta. Si el ancho se agranda a 63 pulg, ¿cuál será la longitud de la pancarta? 84 pulg

5. Carol tiene una fotografía de 24 cm × 36 cm que reduce a $\frac{3}{4}$ de su tamaño. ¿Cuáles son las dimensiones de la nueva fotografía? 18 cm × 27 cm

6. Erik está haciendo un dibujo de la cancha de básquetbol de su escuela. La cancha real mide 84 pies de largo y 50 pies de ancho. Si Erik dibuja la cancha con una longitud de 21 pulg, ¿cuál será el ancho? 12.5 pulg

7. Los cines IMAX tienen las pantallas más grandes del mundo. Hay muchos cines IMAX en todo el mundo. En el museo Henry Ford, situado en Dearborn, Michigan, hay un cine IMAX que tiene una pantalla de 60 pies × 84 pies. Si se cambiara la pantalla de proyección de un salón de clases para que esté en proporción directa con la pantalla IMAX del museo Henry Ford, las dimensiones serían 5 pies × ___ pies. 7

Copyright © by Holt, Rinehart and Winston. All rights reserved.

Copyright © by Holt, Rinehart and Winston. All rights reserved.

LESSON 5-6
Practice
Dilations

Tell whether each transformation is a dilation.

1. 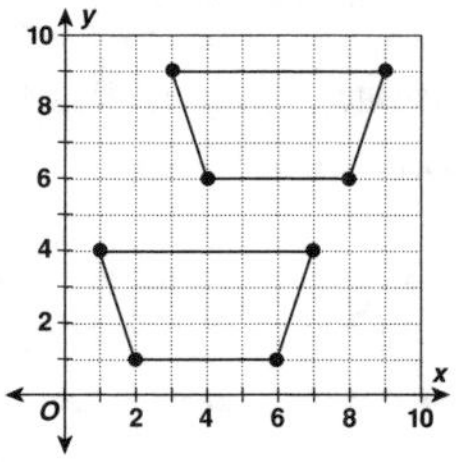

not a dilation

2. 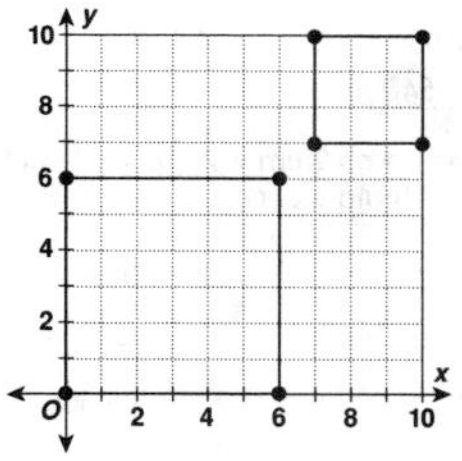

dilation

Dilate each figure by the given scale factor with the origin as the center of dilation. What are the vertices of the image?

3. scale factor of 2

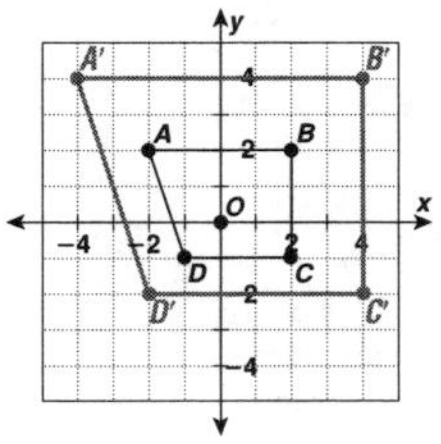

$A'(-4, 4)$, $B'(4, 4)$, $C'(4, -2)$
$D'(-2, -2)$

4. scale factor of $\frac{1}{2}$

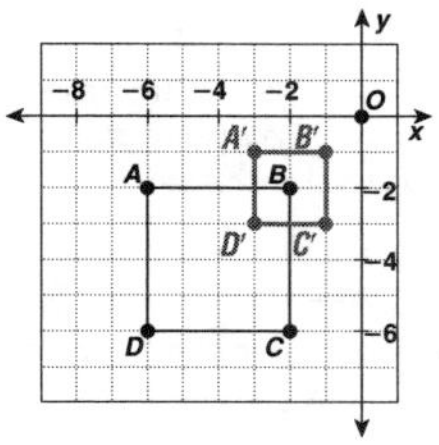

$A'(-3, -1)$, $B'(-1, -1)$
$C'(-1, -3)$, $D'(-3, -3)$

Copyright © by Holt, Rinehart and Winston. All rights reserved.

LECCIÓN 5-6
Práctica
Dilataciones

Indica si cada transformación es una dilatación.

1. 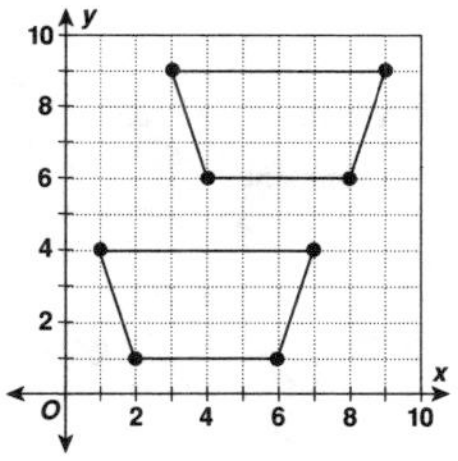

no es una dilatación

2. 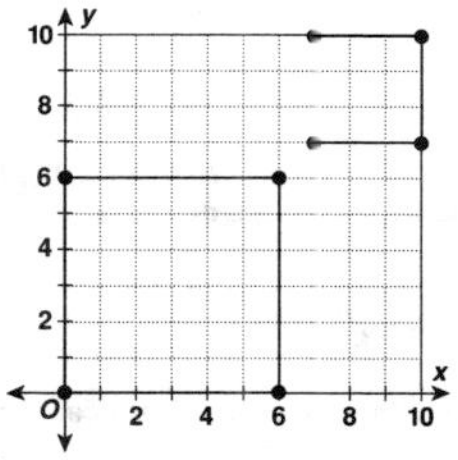

es una dilatación

Dilata cada figura por el factor de escala dado con el origen como centro de dilatación. ¿Cuáles son los vértices de la imagen?

3. factor de escala de 2

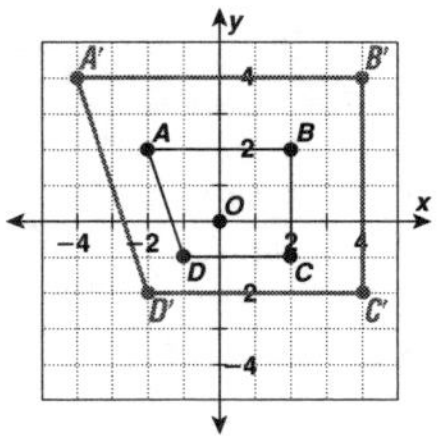

$A'(-4, 4)$, $B'(4, 4)$, $C'(4, -2)$
$D'(-2, -2)$

4. factor de escala de $\frac{1}{2}$

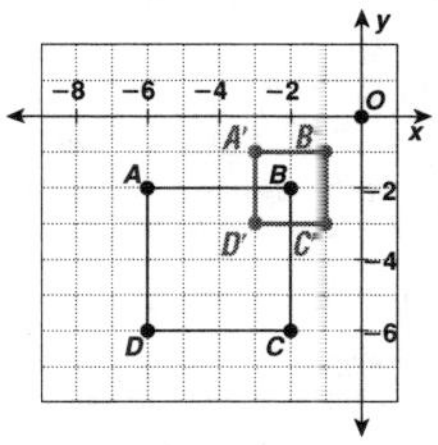

$A'(-3, -1)$, $B'(-1, -1)$
$C'(-1, -3)$, $D'(-3, -3)$

Copyright © by Holt, Rinehart and Winston. All rights reserved.

LESSON 5-7
Practice
Indirect Measurement

1. Tamara wants to know the width of the pond at the park. She drew the diagram and labeled it with the measurements she made. How wide is the pond?

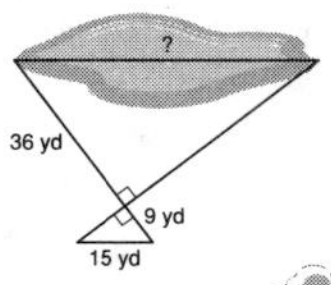

60 yd

Use the diagram for 2 and 3.

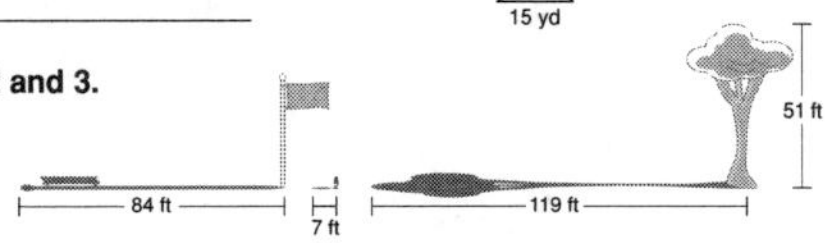

2. How tall is the flagpole?

36 ft

3. How tall is the child?

3 ft

Use the diagram for 4 and 5.

4. How tall is the house?

36 ft

5. The tree is 56 feet tall. How long is its shadow?

112 ft

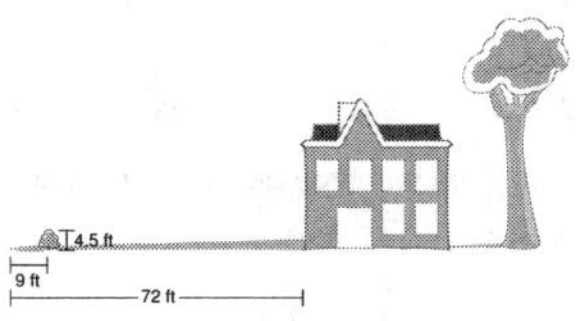

6. Drew wants to know the distance across the river. He drew the diagram and labeled it with the measurements he made. What is the distance across the river?

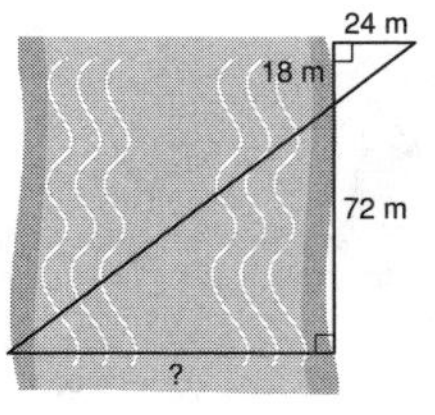

96 m

7. A warehouse is 120 feet tall and casts a shadow 288 feet long. At the same time, Julie casts a shadow 12 feet long. How tall is Julie?

5 ft

Copyright © by Holt, Rinehart and Winston. All rights reserved.

LECCIÓN 5-7
Práctica
Medición indirecta

1. Tamara quiere conocer el ancho de la laguna del parque. Dibujó el diagrama e indicó las mediciones que tomó. ¿Cuál es el ancho de la laguna?

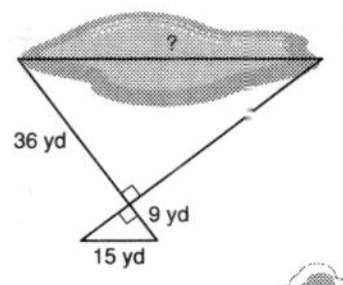

60 yd

Usa el diagrama para los ejercicios 2 y 3.

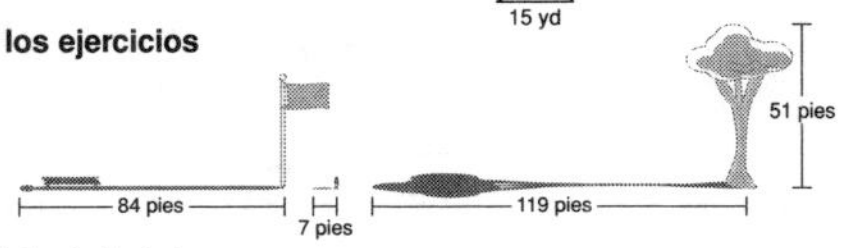

2. ¿Cuál es la altura del mástil de la bandera?

36 pies

3. ¿Cuál es la estatura de niño?

3 pies

Usa el diagrama para los ejercicios 4 y 5.

4. ¿Cuál es la altura de la casa?

36 pies

5. El árbol mide 56 pies de altura. ¿Cuál es la longitud de su sombra?

112 pies

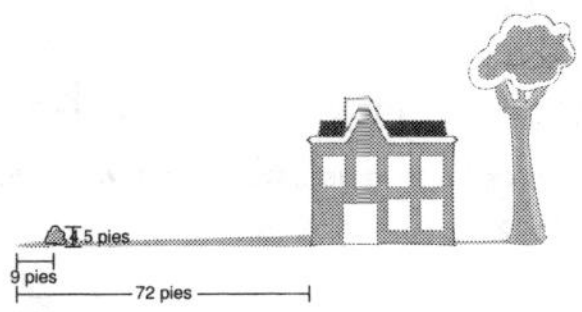

6. Drew quiere saber la distancia que hay entre las orillas del río. Dibujó un diagrama e indicó las mediciones que tomó. ¿Cuál es la distancia entre las orillas del río?

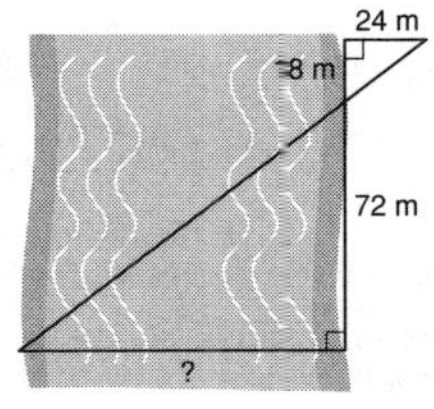

96 m

7. Un depósito mide 120 pies de altura y proyecta una sombra de 288 pies de longitud. En el mismo momento, Julie proyecta una sombra de 12 pies de longitud. ¿Cuál es la estatura de Julie?

5 pies

Copyright © by Holt, Rinehart and Winston. All rights reserved.

Copyright © by Holt, Rinehart and Winston. All rights reserved.

LESSON **5-8**
Practice
Scale Drawings and Scale Models

The scale of a drawing is $\frac{1}{4}$ in. = 15 ft. Find the actual measurement.

1. 9 in. 540 ft
2. 12 in. 720 ft
3. 14 in. 840 ft
4. 15 in. 900 ft

The scale is 2 cm = 25 m. Find the length each measurement would be on a scale drawing.

5. 150 m 12 cm
6. 475 m 38 cm
7. 350 m 28 cm
8. 500 m 40 cm

Tell whether each scale reduces, enlarges, or preserves the size of an actual object.

9. 1 m : 25 cm enlas
10. 8 in. : 1 ft reduces
11. 12 in. : 1 ft serves

12. On a map the distance between Atlanta, Georgia, and Nashville, Tennessee, is 12.5 in. The actual distance between these two cities is 250 miles. What is the scale?

1 in. = 20 mi

13. Blueprints of a house are drawn to the scale of $\frac{1}{4}$ in. = 1 ft. A kitchen measures 3.5 in. by 5 in. on the blueprints. What is the actual size of the kitchen?

14 ft × 20 ft

14. A scale model of a house is 1 ft long. The actual house is 50 ft long. In the model, the window is $1\frac{1}{5}$ in. high. How many feet high is the actual window?

5 ft

15. A model of a skyscraper is 1.6 in. long, 2.8 in. wide, and 11.2 in. high. The scale factor is 8 in. : 250 ft. What are the actual dimensions of the skyscraper?

50 ft loh

Copyright © by Holt, Rinehart and Winston. All rights reserved.

LECCIÓN **5-8**
Práctica
Dibujos y modelos a escala

La escala de un dibujo es $\frac{1}{4}$ pulg = 15 pies. Halla la medida real del dibujo.

1. 9 pulg 540 pies
2. 12 pulg 720 pies
3. 14 pulg 840 pies
4. 15 pulg 900 pies

La escala es 2 cm = 25 m. Halla la longitud que tendría cada medida en un dibujo a escala..

5. 150 m 12 cm
6. 475 m 38 cm
7. 350 m 28 cm
8. 500 m 40 cm

Indica si cada escala reduce, agranda o conserva el tamaño del objeto real.

9. 1 m : 25 cm agranda
10. 8 pulg : 1 pie reduce
11. 12 pulg. : 1 pie conserva

12. En un mapa, la distancia entre Atlanta, Georgia y Nashville, Tennessee, es de 12.5 pulg. La distancia real entre estas dos ciudades es de 250 millas. ¿Cuál es la escala?

1 pulg = 20 mi

13. Se dibuja el plano de una casa a una escala de $\frac{1}{4}$ pulg = 1 pie. La cocina mide 3.5 pulg por 5 pulg en el plano. ¿Cuál es el tamaño real de la cocina?

14 pies × 20 pies

14. El modelo a escala de una casa mide 1 pie de largo. La casa real tiene 50 pies de longitud. En el modelo, la ventana mide $1\frac{1}{5}$ pulg de alto. ¿Cuántas pulgadas de alto mide la ventana real?

5 pies

15. El modelo de un rascacielos mide 1.6 pulg de largo, 2.8 pulg de ancho y 11.2 pulg de alto. El factor de escala es 8 pulg: 250 pies. ¿Cuáles son las dimensiones reales del rascacielos?

50 pies

Copyright © by Holt, Rinehart and Winston.

LESSON **6-1**
Practice
Relating Decimals, Fractions, and Percents

Find the missing ratio or percent equivalent for each letter on the number line.

0%	a	22%	b	r	56%	64%	70%	d	100%
0	$\frac{6}{100}$	m	$\frac{9}{25}$	$\frac{9}{20}$	t	c	x	$\frac{4}{5}$	1

1. *a* 6%
2. *b* 36%
3. *c* $\frac{16}{25}$
4. *d* 80%
5. *m* $\frac{11}{50}$
6. *r* 45%
7. *t* $\frac{14}{25}$
8. *x* $\frac{7}{10}$

Compare. Write <, >, or =.

9. $\frac{3}{4}$ [>] 70%
10. 60% [=] $\frac{3}{5}$
11. 58% [<] 0.6
12. 0.09 [<] 15%
13. $\frac{2}{3}$ [>] 59%
14. 0.45 [>] 40.5%

Order the numbers from least to greatest.

15. 99%, 0.95, $\frac{5}{9}$, 9.5%

9.5%, $\frac{5}{9}$, 0.95, 99%

16. $\frac{3}{8}$, 50%, 0.35, 38%

0.35, $\frac{3}{8}$, 38%, 50%

17. $\frac{4}{5}$, 54%, 0.45, 44.5%

44.5%, 0.45, 54%, $\frac{4}{5}$

18. $\frac{1}{3}$, 20%, 0.3, 3%

3%, 20%, 0.3, $\frac{1}{3}$

19. There are 25 students in math class. Yesterday, 6 students were absent. What percent of the students were absent? 24%

20. Albert spends 2 hours a day on his homework and an hour playing video games. What percent of the day is this? 12.5%

21. Ragu ran the first 3 miles of a 5 mile race in 24 minutes. What percent of the race has he run? 60%

Copyright © by Holt, Rinehart and Winston. All rights reserved.

LECCIÓN **6-1**
Práctica
Cómo relacionar decimales, fracciones y porcentajes

Halla la razón o porcentaje equivalente que falta para cada letra de la recta numérica.

0%	a	22%	b	r	56%	64%	70%	d	100%
0	$\frac{6}{100}$	m	$\frac{9}{25}$	$\frac{9}{20}$	t	c	x	$\frac{4}{5}$	1

1. *a* 6%
2. *b* 36%
3. *c* $\frac{16}{25}$
4. *d* 80%
5. *m* $\frac{11}{50}$
6. *r* 45%
7. *t* $\frac{14}{25}$
8. *x* $\frac{7}{10}$

Compara. Escribe <, > ó =.

9. $\frac{3}{4}$ [>] 70%
10. 60% [=] $\frac{3}{5}$
11. 58% [<] 0.6
12. 0.09 [<] 15%
13. $\frac{2}{3}$ [>] 59%
14. 0.45 [>] 40.5%

Ordena los números de menor a mayor.

15. 99%, 0.95, $\frac{5}{9}$, 9.5%

9.5%, $\frac{5}{9}$, 0.95, 99%

16. $\frac{3}{8}$, 50%, 0.35, 38%

0.35, $\frac{3}{8}$, 38%, 50%

17. $\frac{4}{5}$, 54%, 0.45, 44.5%

44.5%, 0.45, 54%, $\frac{4}{5}$

18. $\frac{1}{3}$, 20%, 0.3, 3%

3%, 20%, 0.3, $\frac{1}{3}$

19. En la clase de matemáticas hay 25 estudiantes. Ayer faltaron 6 estudiantes. ¿Qué porcentaje de los estudiantes faltó? 24%

20. Albert pasa dos horas al día haciendo la tarea y una hora jugando a los videojuegos. ¿Qué porcentaje del día representa esto? 12.5%

21. Ragu corrió las primeras 3 millas de una carrera de 5 millas en 24 minutos. ¿Qué porcentaje de la carrera ha corrido? 60%

Copyright © by Holt, Rinehart and Winston. All rights reserved.

Copyright © by Holt, Rinehart and Winston. All rights reserved.

LESSON 6-2

Practice

Estimate with Percents

Estimate. Estimates may vary.

1. 74% of 99 — about 75
2. 25% of 39 — about 10
3. 52% of 10 — about 5
4. 21% of 50 — about 10
5. 30% of 61 — about 20
6. 24% of 48 — about 12
7. 5% of 41 — about 2
8. 50% of 178 — about 90
9. 33% out of 62 — about 20

Estimate.

10. 48% of 30 is about what number? — about 15
11. 26% of 36 is about what number? — about 9
12. 30% of 22 is about what number? — about 7
13. 21% of 63 is about what number? — about 12
14. Rodney's weekly gross pay is $91. He must pay about 32% in taxes and deductions. Estimate Rodney's weekly take-home pay after deductions. — about $60
15. In the last school election, 492 students voted. Mary received 48% of the votes. About how many votes did she receive? — about 250
16. A restaurant bill for lunch is $14.10. Grace wants to leave a 15% tip and the sales tax rate is 5.5%. About how much will lunch cost Grace in all? — about $17.10
17. A company has found that on average about 6% of the batteries they manufacture are defective. Out of 1,385 batteries, the supervisor assumes that about 83 are defective. Estimate to determine if the manager's number is reasonable? Explain. Yes; 6% of 1,385 is about 5% of 1,400; 10% of 1,400 is 140, and half of 140 is 70. Since 5% of 1,400 is 83 and 70 is close to 70, the estimate is reasonable.

Copyright © by Holt, Rinehart and Winston. All rights reserved.

LECCIÓN 6-2

Práctica

Estimar con porcentajes

Estima. Las estimaciones variarán.

1. 74% de 99 — aproximadamente 75
2. 25% de 39 — aproximadamente 10
3. 52% de 10 — aproximadamente 5
4. 21% de 50 — aproximadamente 10
5. 30% de 61 — aproximadamente 20
6. 24% de 48 — aproximadamente 12
7. 5% de 41 — aproximadamente 2
8. 50% de 178 — aproximadamente 90
9. 33% de 62 — aproximadamente 20

Estima.

10. ¿Aproximadamente cuánto es el 48% de 30? — aproximadamente 15
11. ¿Aproximadamente cuánto es el 26% de 36? — aproximadamente 9
12. ¿Aproximadamente cuánto es el 30% de 22? — aproximadamente 7
13. ¿Aproximadamente cuánto es el 21% de 63? — aproximadamente 12
14. El salario semanal bruto de Rodney es $91. Debe pagar alrededor del 32% en concepto de impuestos y retenciones. Estima el salario semanal neto de Rodney después de las retenciones. — alrededor de $60
15. En la última elección que se llevó a cabo en la escuela votaron 492 estudiantes. Mary obtuvo 48% de los votos. ¿Aproximadamente cuántos votos recibió? — aproximadamente 250
16. La cuenta del almuerzo en un restaurante es de $14.10. Grace quiere dejar un 15% de propina y la tasa de impuesto sobre la venta es del 5.5%. ¿Aproximadamente cuánto le costará el almuerzo en total? — aproximadamente $17.10
17. Una empresa descubrió que, en promedio, alrededor del 6% de las baterías que fabrica tienen fallas. De 1,385 baterías, el gerente calcula que aproximadamente 83 tienen fallas. Haz una estimación para determinar si el cálculo del gerente es razonable. Explica. Sí; el 6 % de 1,385 es aproximadamente 5% de 1,400; el 10% de 1,400 es 140 y la mitad de 140 es 70. Entonces, el 5% de 1,400 es 83 y 83 se aproxima a 70. La estimación es razonable.

Copyright © by Holt, Rinehart and Winston. All rights reserved.

LESSON 6-3

Practice

Finding Percents

Find each percent.

1. What percent of 84 is 21? — 25%
2. 24 is what percent of 60? — 40%
3. What percent of 150 is 75? — 50%
4. What percent of 80 is 68? — 85%
5. 36 is what percent of 80? — 45%
6. What percent of 88 is 33? — 37.5%
7. 19 is what percent of 95? — 20%
8. 28.8 is what percent of 120? — 24%
9. What percent of 56 is 49? — 87.5%
10. What percent of 102 is 17? — $16\frac{2}{3}$
11. What percent of 94 is 42.3? — 45%
12. 90 is what percent of 75? — 120%
13. Daphne bought a used car for $9200. She made a down payment of $1840. Find the percent of the purchase price that is the down payment. — 20%
14. Tricia read $\frac{1}{4}$ of her book on Monday. On Tuesday, she read 36% of the book. On Wednesday, she read 0.27 of the book. She finished the book on Thursday. What percent of the book did she read on Thursday? — 12%
15. An airplane traveled from Boston to Las Vegas making a stop in St. Louis. The plane traveled 2410 miles altogether, which is 230% of the distance from Boston to St. Louis. Find the distance from Boston to St. Louis to the nearest mile. — 1048 mi
16. The first social studies test had 16 questions. The second test had 220% as many questions as the first test. Find the number of questions on the second test. — 36 questions

Copyright © by Holt, Rinehart and Winston. All rights reserved.

LECCIÓN 6-3

Práctica

Cómo hallar porcentajes

Halla cada porcentaje.

1. ¿Qué porcentaje de 84 es 21? — 25%
2. ¿Qué porcentaje de 60 es 24? — 40%
3. ¿Qué porcentaje de 150 es 75? — 50%
4. ¿Qué porcentaje de 80 es 68? — 85%
5. ¿Qué porcentaje de 80 es 36? — 45%
6. ¿Qué porcentaje de 88 es 33? — 37.5%
7. ¿Qué porcentaje de 95 es 19? — 20%
8. ¿Qué porcentaje de 120 es 28.8? — 24%
9. ¿Qué porcentaje de 56 es 49? — 87.5%
10. ¿Qué porcentaje de 102 es 17? — $16\frac{2}{3}$
11. ¿Qué porcentaje de 94 es 42.3? — 45%
12. ¿Qué porcentaje de 75 es 90? — 120%
13. Daphne compró un automóvil usado a $9200. Hizo un pago inicial de $1840. Halla el porcentaje del precio de compra que representa el pago inicial. — 20%
14. Tricia leyó $\frac{1}{4}$ de su libro el lunes. El martes, leyó 36% del libro. El miércoles, leyó 0.27 del libro. Terminó el libro el jueves. ¿Qué porcentaje del libro leyó el jueves? — 12%
15. Un avión viajó de Boston a Las Vegas e hizo una escala en St. Louis. En total, el avión recorrió 2410 millas, que representan el 230% de la distancia entre Boston y St. Louis. Halla la distancia entre Boston y St. Louis a la milla más cercana. — 1048 mi
16. El primer examen de estudios sociales tenía 16 preguntas. El segundo examen tenía 220% de la cantidad de preguntas del primero. Halla la cantidad de preguntas que tenía el segundo examen. — 36 preguntas

Copyright © by Holt, Rinehart and Winston. All rights reserved.

Copyright © by Holt, Rinehart and Winston. All rights reserved.

LESSON 6-4
Practice
Finding a Number When the Percent Is Known

Find each number to the nearest tenth.

1. 40% of what number is 18? 45
2. 28 is 35% of what number? 80
3. 21 is 60% of what number? 35
4. 25% of what number is 19? 76
5. 40% of what number is 22? 55
6. 41 is 50% of what number? 82
7. 50 is 15% of what number? 333.3
8. 0.3% of what number is 24? 8,000
9. 36 is 30% of what number? 120
10. 26 is 75% of what number? 34.7
11. 12.5% of what number is 14? 112
12. 25% of what number is 28.25? 113
13. 27 is $33\frac{1}{3}$% of what number? 81
14. 54 is 150% of what number? 36
15. There were 546 students at a school assembly. This was 65% of all students who attend Content Middle School. How many students attend Content Middle School?
    840 students
16. On his last test Greg answered 64 questions correctly. This was 80% of the questions. How many questions were on the test?
    80 questions
17. The price of a jacket is $48. If the sales tax rate is 5.5%, what is the amount of tax? What is the total cost of the jacket?
    tax is $2.64; total cost is $50.64
18. Carla has finished swimming 14 laps in swim practice. This is 70% of the total number of laps she must swim. How many more laps must Carla swim to complete her practice?
    6 laps

Copyright © by Holt, Rinehart and Winston. All rights reserved.

LECCIÓN 6-4
Práctica
Cómo hallar un número cuando se conoce el porcentaje

Halla cada número a la décima más cercana.

1. ¿El 40% de qué número es 18? 45
2. ¿El 35% de qué número es 28? 80
3. ¿El 60% de qué número es 21? 35
4. ¿El 25% de qué número es 19? 76
5. ¿El 40% de qué número es 22? 55
6. ¿El 50% de qué número es 41? 82
7. ¿El 15% de qué número es 50? 333.3
8. ¿El 0.3% de qué número es 24? 8,000
9. ¿El 30% de qué número es 36? 120
10. ¿El 75% de qué número es 26? 34.7
11. ¿El 12.5% de qué número es 14? 112
12. ¿El 25% de qué número es 28.25? 113
13. ¿El $33\frac{1}{3}$% de qué número es 27? 81
14. ¿El 150% de qué número es 54? 36
15. En una asamblea de la escuela había 546 estudiantes. Esto representa el 65% de todos los estudiantes que asisten a la escuela intermedia Content. ¿Cuántos estudiantes asisten a la escuela intermedia Content?
    840 estudiantes
16. En el último examen, Greg respondió correctamente 64 preguntas. Esto equivale al 80% de las preguntas. ¿Cuántas preguntas tenía el examen?
    80 preguntas
17. Una chaqueta cuesta $48. Si la tasa de impuesto sobre la venta es del 5.5%, ¿cuál es la cantidad del impuesto? ¿Cuál es el costo total de la chaqueta?
    el impuesto es de $2.64; el costo total es de $50.64
18. Carla ha completado 14 vueltas en el entrenamiento de natación. Esto equivale al 70% del total de vueltas que debe nadar. ¿Cuántas vueltas más debe nadar para terminar el entrenamiento?
    6 vueltas

Copyright © by Holt, Rinehart and Winston. All rights reserved.

LESSON 6-5
Practice
Percent Increase and Decrease

Find each percent increase or decrease to the nearest percent.

1. from 16 to 20 — increase 25%
2. from 30 to 24 — decrease 20%
3. from 15 to 30 — increase 100%
4. from 35 to 21 — decrease 40%
5. from 40 to 46 — increase 15%
6. from 45 to 63 — increase 40%
7. from 18 to 26.1 — increase 45%
8. from 24.5 to 21.56 — decrease 12%
9. from 90 to 72 — decrease 20%
10. from 29 to 54 — increase 86%
11. from 42 to 92.4 — increase 120%
12. from 38 to 33 — decrease 13%
13. from 64 to 36.4 — decrease 43%
14. from 78 to 136.5 — increase 75%
15. from 89 to 32.9 — decrease 63%
16. Mr. Havel bought a car for $2400 and sold it for $2700. What was the percent of profit for Mr. Havel in selling the car? 12.5%
17. A computer store buys a computer program for $24 and sells it for $91.20. What is the percent of increase in the price? 280%
18. A manufacturing company with 450 employees begins a new product line and must add 81 more employees. What is the percent of increase in the number of employees? 18%
19. Richard earns $2700 a month. He received a 3% raise. What is Richard's new annual salary? $33,372
20. Marlis has 765 cards in her baseball card collection. She sells 153 of the cards. What is the percent of decrease in the number of cards in the collection? 20%

Copyright © by Holt, Rinehart and Winston. All rights reserved.

LECCIÓN 6-5
Práctica
Porcentaje de incremento y disminución

Halla cada porcentaje de incremento o disminución al porcentaje más cercano.

1. de 16 a 20 — 25% de incremento
2. de 30 a 24 — 20% de disminución
3. de 15 a 30 — 100% de incremento
4. de 35 a 21 — 40% de disminución
5. de 40 a 46 — 15% de incremento
6. de 45 a 63 — 40% de incremento
7. de 18 a 26.1 — 45% de incremento
8. de 24.5 a 21.56 — 12% de disminución
9. de 90 a 72 — 20% de disminución
10. de 29 a 54 — 86% de incremento
11. de 42 a 92.4 — 120% de incremento
12. de 38 a 33 — 3% de disminución
13. de 64 a 36.4 — 43% de disminución
14. de 78 a 136.5 — 75% de incremento
15. de 89 a 32.9 — 63% de disminución
16. El señor Havel compró un auto a $2400 y lo vendió a $2700. ¿Cuál fue el porcentaje de ganancia que obtuvo el señor Havel al vender el auto? 12.5%
17. Una tienda de computación compra un programa de computación a $24 y lo vende a $91.20. ¿Cuál es el porcentaje de incremento en el precio? 280%
18. Una compañía fabricante que tiene 450 empleados comienza a producir una nueva línea de productos y debe agregar 81 empleados más. ¿Cuál es el porcentaje de incremento en el número de empleados? 18%
19. Richard gana $2700 por mes. Recibió un aumento del 3%. ¿Cuál es el nuevo salario anual de Richard? $33,372
20. Marlis tiene 765 tarjetas en la colección de tarjetas de béisbol. Vende 153 tarjetas de su colección. ¿Cuál es el porcentaje de disminución en la cantidad de tarjetas que tiene en su colección? 20%

Copyright © by Holt, Rinehart and Winston. All rights reserved.

Copyright © by Holt, Rinehart and Winston. All rights reserved.

LESSON **6-6**
Practice
Applications of Percents

Complete the table to find the amount of sales tax for each sale amount to the nearest cent.

1.

Sale amount	5% sales tax	8% sales tax	6.5% sales tax
$67.50	$3.38	$5.40	$4.39
$98.75	$4.94	$7.90	$6.42
$399.79	$19.99	$31.98	$25.99
$1250.00	$62.50	$100.00	$81.25

Complete the table to find the commission for each sale amount to the nearest cent.

2.

Sale amount	6% commision	9% commision	8.5% commission
$475.00	$28.50	$42.75	$40.38
$2450.00	$147.00	$220.50	$208.25
$12,500.00	$750.00	$1125.00	$1062.50
$98,900.00	$5934.00	$8901.00	$8406.50

3. Alice earns a monthly salary of $315 plus a commission on her total sales. Last month her total sales were $9640, and she earned a total of $1182.60. What is her commission rate? 9%

4. Phillipe works for a computer store that pays a 12% commission and no salary. What will Phillipe's weekly sales have to be for him to earn $360? $3000

5. The purchase price of a book is $35.85. The sales tax rate is 6.5%. How much is the sales tax to the nearest cent? What is the total cost of the book?

sales tax is $2.33; total cost is $38.18

6. Who made more commission this month? How much did she make? Salesperson A made 11% of $67,530. Salesperson B made 8% of $85,740.

Salesperson A $7428.30

7. Jon earned $38,000 last year. He paid $6,840 towards entertainment. What percent of his earnings did Jon pay in entertainment expenses? 18%

8. The Cougars won 62% of their games. They won 93 games. How many games did they lose? 57 games

Copyright © by Holt, Rinehart and Winston.
All rights reserved.

LECCIÓN **6-6**
Práctica
Aplicaciones de porcentajes

Completa la tabla para hallar la cantidad del impuesto sobre la venta para cada cantidad de venta al centavo más cercano.

1.

Cantidad de la venta	5% de impuesto sobre la venta	8% de impuesto sobre la venta	6.5% de impuesto sobre la venta
$67.50	$3.38	$5.40	$4.39
$98.75	$4.94	$7.90	$6.42
$399.79	$19.99	$31.98	$25.99
$1250.00	$62.50	$100.00	$81.25

Completa la tabla para hallar la comisión por cada venta al centavo más cercano.

2.

Cantidad de la venta	6% de comisión	9% de comisión	8.5% de comisión
$475.00	$28.50	$42.75	$40.38
$2450.00	$147.00	$220.50	$208.25
$12,500.00	$750.00	$1125.00	$1062.50
$98,900.00	$5934.00	$8901.00	$8406.50

3. Alice cobra un sueldo mensual de $315 más una comisión sobre el total de sus ventas. El mes pasado, el total de sus ventas fue de $9640 y ganó un total de $1182.60. ¿Cuál es la tasa de comisión de Alice? 9%

4. Phillipe trabaja en una tienda de computación que paga el 12% de comisión, sin salario. ¿Cuánto debe vender Phillipe por semana para ganar $360? $3000

5. El precio de compra de un libro es de $35.85. La tasa de impuesto sobre la venta es del 6.5% ¿Cuánto es el impuesto sobre la venta al centavo más cercano? ¿Cuál es el costo total del libro?

el impuesto sobre la venta es de $2.33; el costo total es de $38.18

6. ¿Quién ganó más en concepto de comisiones este mes? ¿Cuánto ganó? La vendedora A ganó el 11% de $67,530. La vendedora B ganó el 8% de $85,740.

la vendedora A, $7428.30

7. Jon ganó $38,000 el año pasado. Gastó $6,840 en entretenimiento. ¿Qué porcentaje de sus ingresos gastó en entretenimiento? 18%

8. Los Pumas ganaron el 62% de sus partidos. Ganaron 93 partidos. ¿Cuántos partidos perdieron? 57 partidos

Copyright © by Holt, Rinehart and Winston.
All rights reserved.

LESSON **6-7**
Practice
Simple Interest

Find the missing value.

1. principal = $125
rate = 4%
time = 2 years
interest = ?
$10

2. principal = ?
rate = 5%
time = 4 years
interest = $90
$450

3. principal = $150
rate = 6%
time = ? years
interest = $54
6 years

4. principal = $200
rate = ?%
time = 3 years
interest = $30
5%

5. principal = $550
rate = ?%
time = 3 years
interest = $57.75
3.5%

6. principal = ?
rate = $3\frac{1}{4}$%
time = 2 years
interest = $63.05
$970

7. Kwang deposits money in an account that earns 5% simple interest. He earned $546 in interest 2 years later. How much did he deposit? $5460

8. Simon opened a certificate of deposit with the money from his bonus check. The bank offered 4.5% interest for 3 years of deposit. Simon calculated that he would earn $87.75 interest in that time. How much did Simon deposit to open the account? $650

9. Douglas borrowed $1000 from Patricia. He agreed to repay her $1150 after 3 years. What was the interest rate of the loan? 5%

10. What is the interest paid for a loan of $800 at 5% annual interest for 9 months? $30

Copyright © by Holt, Rinehart and Winston.
All rights reserved.

LECCIÓN **6-7**
Práctica
Interés simple

Halla el valor que falta.

1. capital = $125
tasa = 4%
tiempo = 2 años
interés = ?
$10

2. capital = ?
tasa = 5%
tiempo = 4 años
interés = $90
$450

3. capital = $150
tasa = 6%
tiempo = ? años
interés = $54
6 años

4. capital = $200
tasa = ?%
tiempo = 3 años
interés = $30
5%

5. capital = $550
tasa = ?%
tiempo = 3 años
interés = $57.75
3.5%

6. capital = ?
tasa = $3\frac{1}{4}$%
tiempo = 2 años
interés = $63.05
$970

7. Kwang deposita dinero en una cuenta que le da el 5% de interés simple. Después de dos años, ganó $546 en intereses. ¿Cuánto dinero depositó? $5460

8. Simón abrió un certificado de depósito con el dinero de su cheque de bonificación. El banco ofreció el 4.5% de interés por un depósito a 3 años. Simón calculó que ganaría $87.75 en intereses para ese momento. ¿Cuánto dinero depositó para abrir la cuenta? $650

9. Douglas le pidió un préstamo de $1000 a Patricia. Acordaron que le devolvería $1150 en 3 años. ¿Cuál era la tasa de interés del préstamo? 5%

10. ¿Cuál es el interés que se pagó por un préstamo de $800 a una tasa de interés anual del 5% por 9 meses? $30

Copyright © by Holt, Rinehart and Winston.
All rights reserved.

Copyright © by Holt, Rinehart and Winston.
All rights reserved.

LESSON 7-1
Practice
Points, Lines, Planes, and Angles

Use the diagram to name each figure.

1. four points

   R, S, T, W

2. a line

   $\overleftrightarrow{WT}$

3. a plane

   Possible answer: *RST*

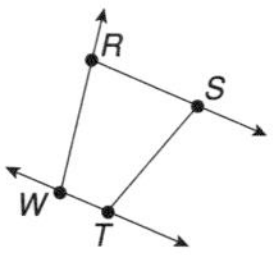

4. three segments

   $\overline{RS}$, $\overline{ST}$, $\overline{TW}$, or $\overline{RW}$

5. four rays

   $\overrightarrow{RS}$, $\overrightarrow{WT}$, $\overrightarrow{WR}$, $\overrightarrow{TW}$

Use the diagram to name each figure.

6. a right angle

   $\angle NXP$, $\angle MXR$, $\angle MXN$, or $\angle RXP$

7. two acute angles

   $\angle MXS$ and $\angle RXS$

8. two obtuse angles

   $\angle NXS$ and $\angle PXS$

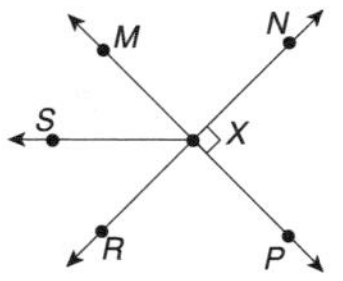

9. a pair of complementary angles

   $\angle RXS$ and $\angle SXM$

10. three pairs of supplementary angles

   Possible answer: $\angle NXP$ and $\angle MXN$, $\angle RXM$ and $\angle MXN$, $\angle RXP$ and $\angle PXN$

In the figure, $\angle 1$ and $\angle 3$ are vertical angles, and $\angle 2$ and $\angle 4$ are vertical angles.

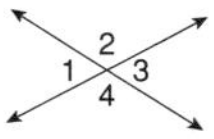

11. If $m\angle 2 = 110°$, find $m\angle 4$.

   $m\angle 4 = 110°$

12. If $m\angle 1 = n°$, find $m\angle 3$.

   $m\angle 3 = n°$

Copyright © by Holt, Rinehart and Winston. All rights reserved.

LECCIÓN 7-1
Práctica
Puntos, líneas, planos y ángulos

Usa el diagrama para identificar cada figura.

1. cuatro puntos

   R, S, T, W

2. una línea

   $\overleftrightarrow{WT}$

3. un plano

   Respuesta posible: *RST*

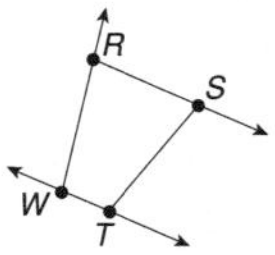

4. tres segmentos

   $\overline{RS}$, $\overline{ST}$, $\overline{TW}$ o $\overline{RW}$

5. cuatro rayos

   $\overrightarrow{RS}$, $\overrightarrow{WT}$, $\overrightarrow{WR}$, $\overrightarrow{TW}$

Usa el diagrama para identificar cada figura.

6. un ángulo recto

   $\angle NXP$, $\angle MXR$, $\angle MXN$ o $\angle RXP$

7. dos ángulos agudos

   $\angle MXS$ y $\angle RXS$

8. dos ángulos obtusos

   $\angle NXS$ y $\angle PXS$

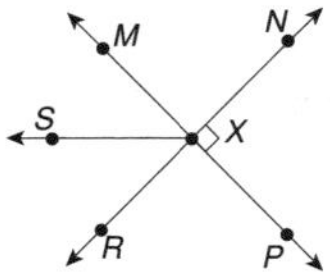

9. un par de ángulos complementarios

   $\angle RXS$ y $\angle SXM$

10. tres pares de ángulos suplementarios

   Respuesta posible: $\angle NXP$ y $\angle MXN$, $\angle RXM$ y $\angle MXN$, $\angle RXP$ y $\angle PXN$

En la figura, $\angle 1$ y $\angle 3$ son ángulos opuestos por el vértice y $\angle 2$ y $\angle 4$ también lo son.

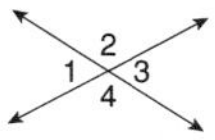

11. Si $m\angle 2 = 110°$, halla $m\angle 4$.

   $m\angle 4 = 110°$

12. Si $m\angle 1 = n°$, halla $m\angle 3$.

   $m\angle 3 = n°$

Copyright © by Holt, Rinehart and Winston. All rights reserved.

LESSON 7-2
Practice
Parallel and Perpendicular Lines

1. Measure the angles formed by the transversal and the parallel lines. Which angles seem to be congruent?

   $\angle 1 \cong \angle 4 \cong \angle 5 \cong \angle 8$ and

   $\angle 2 \cong \angle 3 \cong \angle 6 \cong \angle 7$

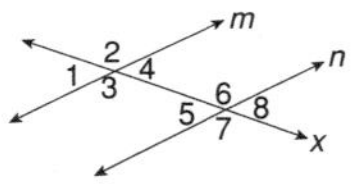

In the figure, line *m* ∥ line *n*. Find the measure of each angle.

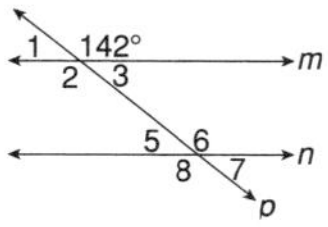

2. $\angle 1$ 38°
3. $\angle 2$ 142°
4. $\angle 5$ 38°
5. $\angle 6$ 142°
6. $\angle 8$ 142°
7. $\angle 7$ 38°

In the figure, line *a* ∥ line *b*. Find the measure of each angle.

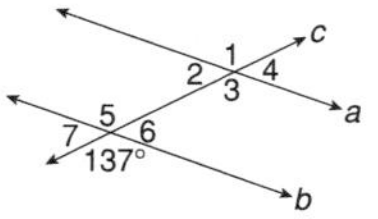

8. $\angle 2$ 43°
9. $\angle 5$ 137°
10. $\angle 6$ 43°
11. $\angle 7$ 43°
12. $\angle 4$ 43°
13. $\angle 3$ 137°

In the figure, line *r* ∥ line *s*.

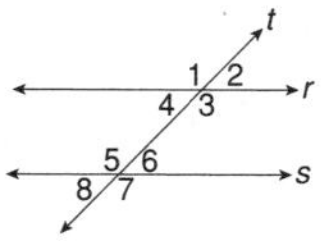

14. Name all angles congruent to $\angle 2$.

   $\angle 4$, $\angle 6$, $\angle 8$

15. Name all angles congruent to $\angle 7$.

   $\angle 1$, $\angle 3$, $\angle 5$

16. Name three pairs of supplementary angles.

   Possible answer: $\angle 1 + \angle 2$, $\angle 3 + \angle 4$, $\angle 5 + \angle 6$, or $\angle 7 + \angle 8$

17. Which line is the transversal?

   line *t*

Copyright © by Holt, Rinehart and Winston. All rights reserved.

LECCIÓN 7-2
Práctica
Líneas paralelas y perpendiculares

1. Mide los ángulos formados por las líneas paralelas y la transversal. ¿Qué angulos parecen ser congruentes?

   $\angle 1 \cong \angle 4 \cong \angle 5 \cong \angle 8$ y

   $\angle 2 \cong \angle 3 \cong \angle 6 \cong \angle 7$

En la figura, línea *m* ∥ línea *n*. Halla la medida de cada ángulo.

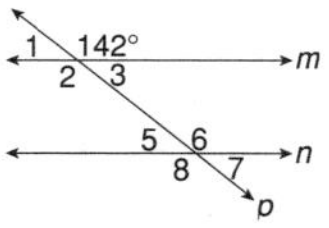

2. $\angle 1$ 38°
3. $\angle 2$ 142°
4. $\angle 5$ 38°
5. $\angle 6$ 142°
6. $\angle 8$ 142°
7. $\angle 7$ 38°

En la figura, línea *a* ∥ línea *b*. Halla la medida de cada ángulo.

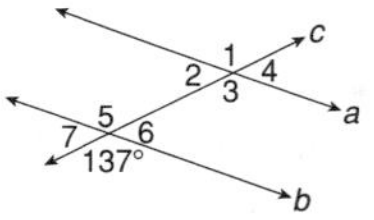

8. $\angle 2$ 43°
9. $\angle 5$ 137°
10. $\angle 6$ 43°
11. $\angle 7$ 43°
12. $\angle 4$ 43°
13. $\angle 3$ 137°

En la figura, línea *r* ∥ línea *s*.

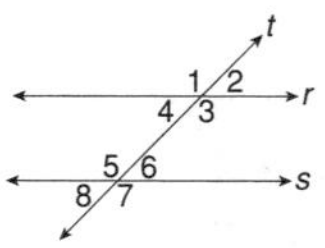

14. Indica todos los ángulos congruentes con $\angle 2$.

   $\angle 4$, $\angle 6$, $\angle 8$

15. Indica todos los ángulos congruentes con $\angle 7$.

   $\angle 1$, $\angle 3$, $\angle 5$

16. Indica tres pares de ángulos suplementarios.

   Respuesta posible: $\angle 1 + \angle 2$, $\angle 3 + \angle 4$, $\angle 5 + \angle 6$, o $\angle 7 + \angle 8$

17. ¿Cuál es la línea transversal?

   la línea *t*

Copyright © by Holt, Rinehart and Winston. All rights reserved.

Copyright © by Holt, Rinehart and Winston. All rights reserved.

LESSON **7-3**

Practice

Angles in Triangles

1. Find $x°$ in the right triangle.
2. Find $y°$ in the obtuse triangle.
3. Find $m°$ in the acute triangle.

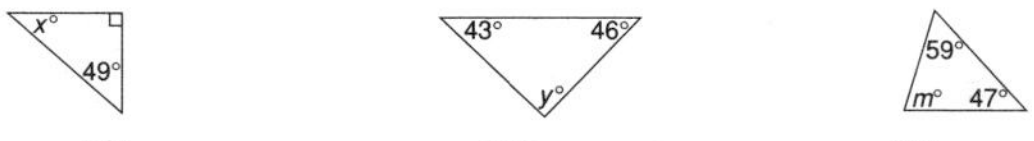

1. 41°
2. 91°
3. 74°

4. Find $n°$ in the obtuse triangle.
5. Find $w°$ in the acute triangle.
6. Find $t°$ in the right triangle.

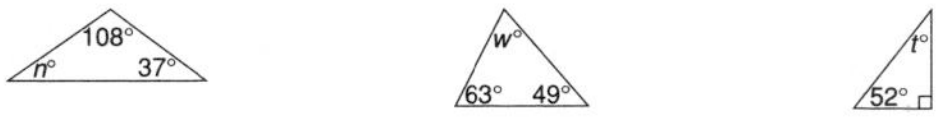

4. 35°
5. 68°
6. 38°

7. Find $t°$ in the scalene triangle.
8. Find $x°$ in the isosceles triangle.
9. Find $n°$ in the scalene triangle.

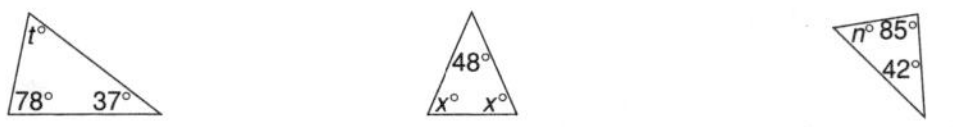

7. 65°
8. 66°
9. 53°

10. Find $x°$ in the isosceles triangle.
11. Find y in the equilateral triangle.
12. Find r in the isoceles triangle.

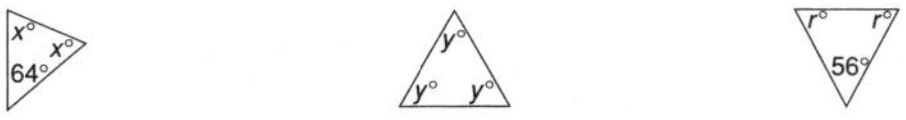

10. 58°
11. 60°
12. 62°

13. The second angle in a triangle is one third as large as the first. The third angle is two thirds as large as the first angle. Find the angle measures. Draw a possible picture of the triangle.

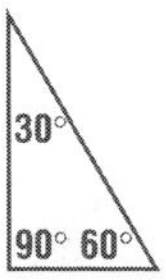

30°, 60°, 90° sample answer

Copyright © by Holt, Rinehart and Winston. All rights reserved.

LECCIÓN **7-3**

Práctica

Ángulos en triángulos

1. Halla $x°$ en el triángulo rectángulo.
2. Halla $y°$ en el triángulo obtusángulo.
3. Halla $m°$ en el triángulo acutángulo.

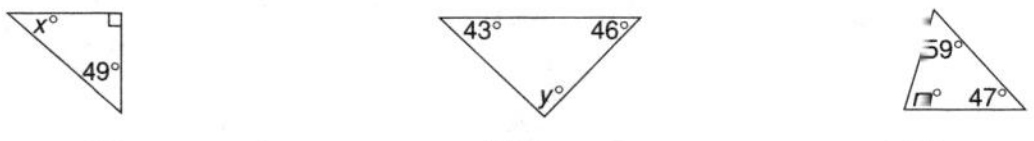

1. 41°
2. 91°
3. 74°

4. Halla $n°$ en el triángulo obtusángulo.
5. Halla $w°$ en el triángulo acutángulo.
6. Halla $t°$ en el triángulo rectángulo.

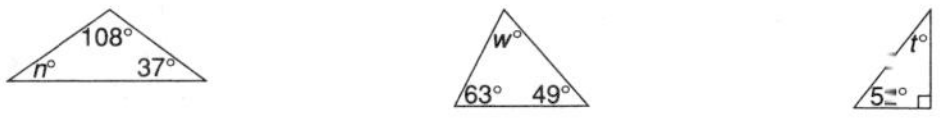

4. 35°
5. 68°
6. 38°

7. Halla $t°$ en el triángulo escaleno.
8. Halla $x°$ en el triángulo isósceles.
9. Halla $n°$ en el triángulo escaleno.

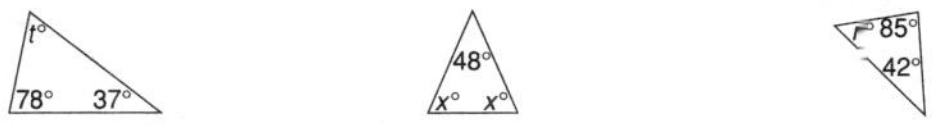

7. 65°
8. 66°
9. 53°

10. Halla $x°$ en el triángulo isósceles.
11. Halla $y°$ en el triángulo equilátero.
12. Halla $r°$ en el triángulo isósceles.

10. 58°
11. 60°
12. 62°

13. El segundo ángulo de un triángulo mide un tercio del tamaño del primero. El tercer ángulo mide dos tercios del tamaño del primero. Halla las medidas de los ángulos y dibuja una figura posible.

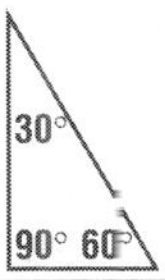

30°, 60°, 90° ejemplo de respuesta

Copyright © by Holt, Rinehart and Winston. All rights reserved.

LESSON **7-4**

Practice

Classifying Polygons

Find the sum of the angle measures in each figure.

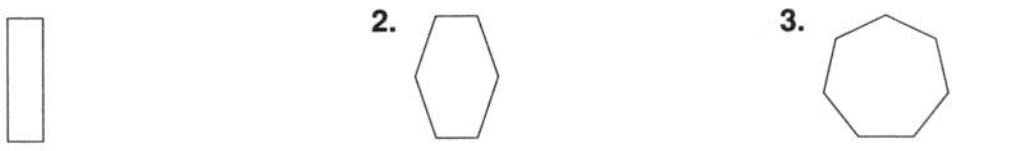

1. 360°
2. 720°
3. 900°

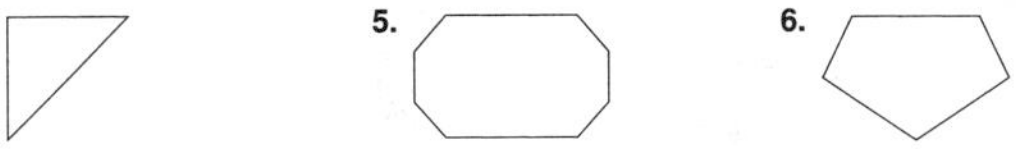

4. 180°
5. 1080°
6. 540°

Find the angle measures in each regular polygon.

7. 120°
8. 90°
9. 135°

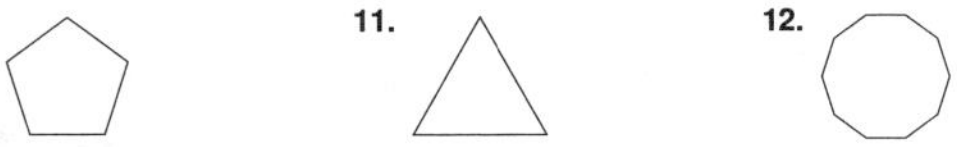

10. 108°
11. 60°
12. 144°

Give all the names that apply to each figure.

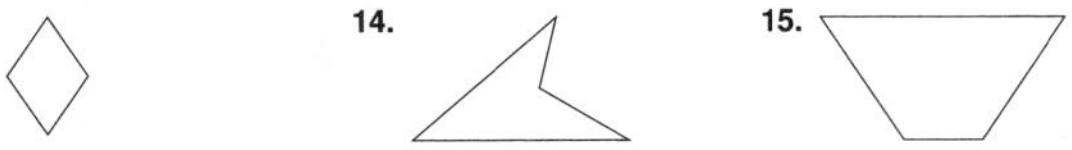

13. quadrilateral, parallelogram, rhombus
14. quadrilateral
15. quadrilateral, trapezoid

Copyright © by Holt, Rinehart and Winston. All rights reserved.

LECCIÓN **7-4**

Práctica

Cómo clasificar polígonos

Halla la suma de las medidas de los ángulos de cada figura.

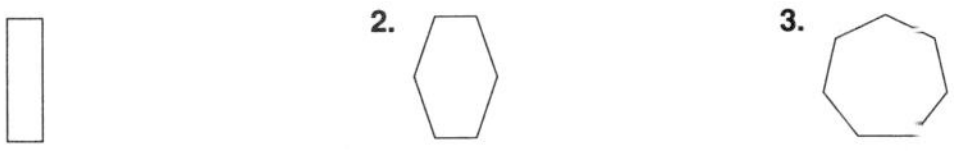

1. 360°
2. 720°
3. 900°

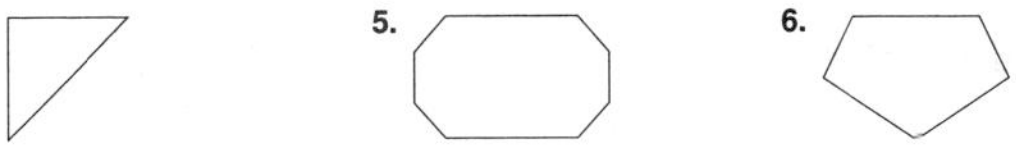

4. 180°
5. 1080°
6. 540°

Halla las medidas de los ángulos de cada polígono regular.

7. 120°
8. 90°
9. 135°

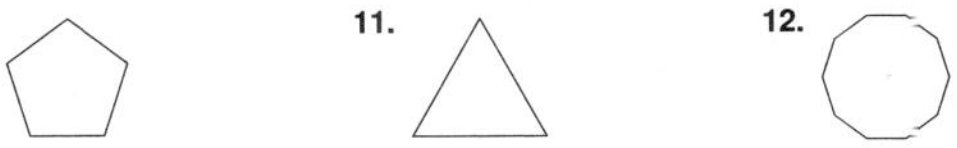

10. 108°
11. 60°
12. 144°

Menciona todos los nombres que correspondan a cada figura.

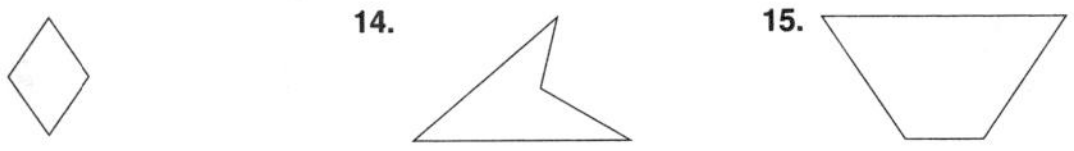

13. cuadrilátero, paralelogramo, rombo
14. cuadrilátero
15. cuadrilátero, trapecio

Copyright © by Holt, Rinehart and Winston. All rights reserved.

Copyright © by Holt, Rinehart and Winston. All rights reserved.

LESSON **7-5**

Practice
Coordinate Geometry

Determine if the slope of each line is positive, negative, 0, or undefined. Then find the slope of each line.

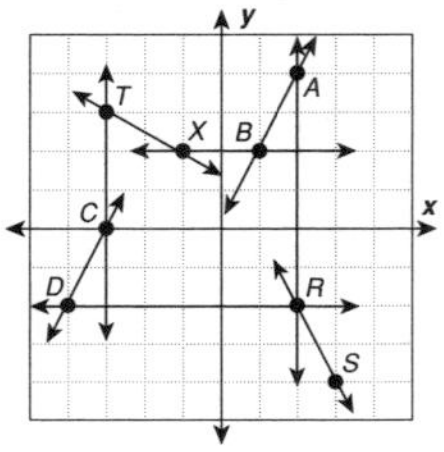

1. $\overleftrightarrow{AB}$

   positive; 2

2. $\overleftrightarrow{CD}$

   positive; 2

3. $\overleftrightarrow{RS}$

   negative; −2

4. $\overleftrightarrow{TC}$

   undefined

5. $\overleftrightarrow{DR}$

   0

6. $\overleftrightarrow{TX}$

   negative; $-\frac{1}{2}$

7. Which lines are parallel?

   $\overleftrightarrow{AB} \parallel \overleftrightarrow{CD}$; $\overleftrightarrow{TC} \parallel \overleftrightarrow{AR}$; $\overleftrightarrow{XB} \parallel \overleftrightarrow{DR}$

8. Which lines are perpendicular?

   $\overleftrightarrow{AB} \perp \overleftrightarrow{TX}$; $\overleftrightarrow{CD} \perp \overleftrightarrow{TX}$; $\overleftrightarrow{DR} \perp \overleftrightarrow{TC}$; $\overleftrightarrow{DR} \perp \overleftrightarrow{AR}$; $\overleftrightarrow{TC} \perp \overleftrightarrow{XB}$; $\overleftrightarrow{AR} \perp \overleftrightarrow{XB}$

Graph the quadrilateral with the given vertices. Write all the names that apply to the quadrilateral.

9. (−1, 1), (4, 1), (1, −3), (−4, −3)

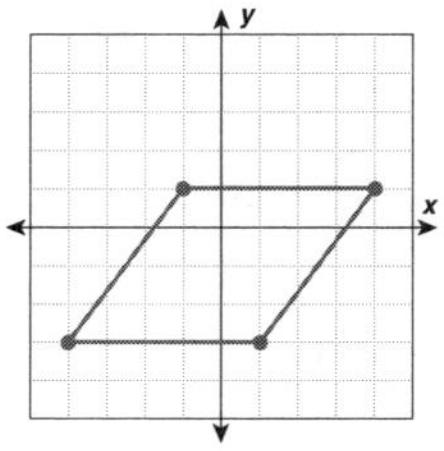

quadrilateral, parallelogram,

rhombus

Find the coordinates of the missing vertex.

10. rhombus *ABCD* with *A*(0, 4), *B*(4, 1), and *C*(0, −2)

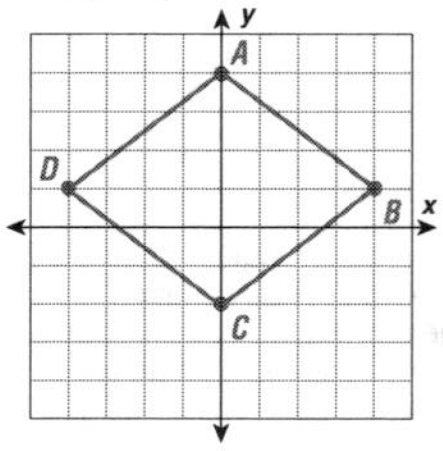

D (−4, 1)

Copyright © by Holt, Rinehart and Winston.
All rights reserved.

LECCIÓN **7-5**

Práctica
Geometría de coordenadas

Determina si la pendiente de cada línea es positiva, negativa, 0 o indefinida. Luego, halla la pendiente de cada línea.

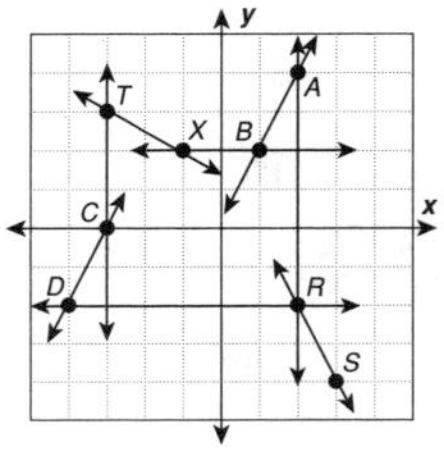

1. $\overleftrightarrow{AB}$

   positiva; 2

2. $\overleftrightarrow{CD}$

   positiva; 2

3. $\overleftrightarrow{RS}$

   negativa; −2

4. $\overleftrightarrow{TC}$

   indefinida

5. $\overleftrightarrow{DR}$

   0

6. $\overleftrightarrow{TX}$

   negativa; $-\frac{1}{2}$

7. ¿Qué líneas son paralelas?

   $\overleftrightarrow{AB} \parallel \overleftrightarrow{CD}$; $\overleftrightarrow{TC} \parallel \overleftrightarrow{AR}$; $\overleftrightarrow{XB} \parallel \overleftrightarrow{DR}$

8. ¿Qué líneas son perpendiculares?

   $\overleftrightarrow{AB} \perp \overleftrightarrow{TX}$; $\overleftrightarrow{CD} \perp \overleftrightarrow{TX}$; $\overleftrightarrow{DR} \perp \overleftrightarrow{TC}$; $\overleftrightarrow{DR} \perp \overleftrightarrow{AR}$; $\overleftrightarrow{TC} \perp \overleftrightarrow{XB}$; $\overleftrightarrow{AR} \perp \overleftrightarrow{XB}$

Representa gráficamente el cuadrilátero con los vértices dados. Menciona todos los nombres que correspondan al cuadrilátero.

9. (−1, 1), (4, 1), (1, −3), (−4, −3)

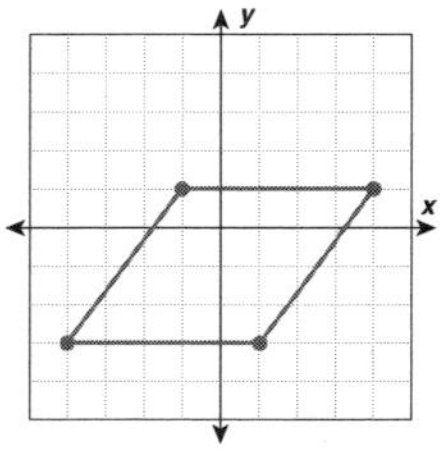

cuadrilátero, paralelogramo,

rombo

Halla las coordenadas del vértice que falta.

10. rombo *ABCD* con *A*(0, 4), *B*(4, 1), y *C*(0, −2)

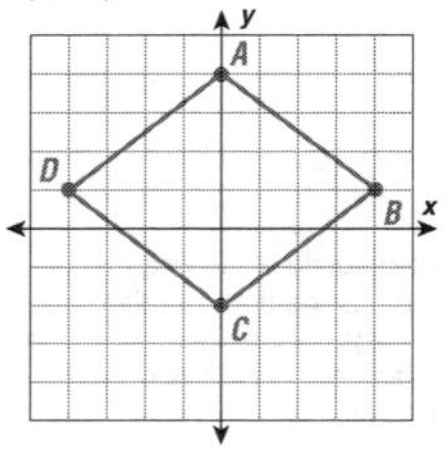

D (−4, 1)

Copyright © by Holt, Rinehart and Winston.
All rights reserved.

LESSON **7-6**

Practice
Congruence

Write a congruence statement for each pair of polygons.

1.

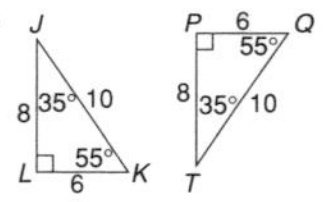

triangle *JKL* ≅ triangle *TQP*

2.

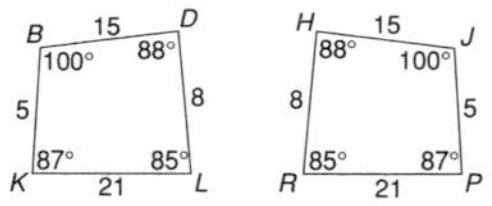

quadrilateral *BDLK* ≅

quadrilateral *JHRP*

3.

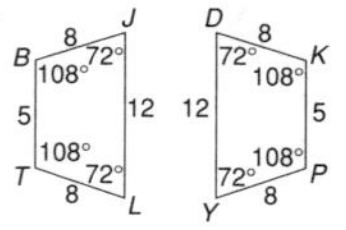

trapezoid *BJLT* ≅

trapezoid *KDYP*

4.

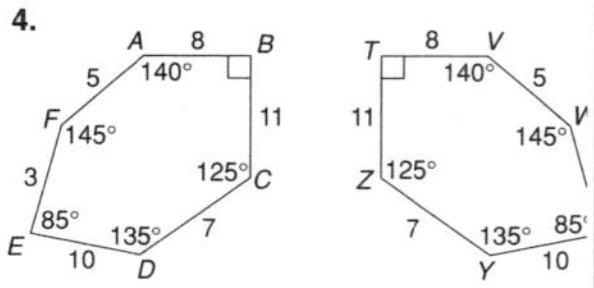

hexagon *ABCDEF* ≅

hexagon *VTZYXW*

In the figure, triangle *PRT* ≅ triangle *FJH*.

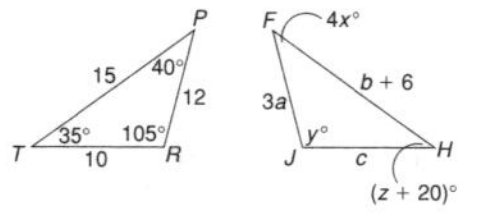

5. Find *a*.

   4

6. Find *b*.

   9

7. Find *c*.

   10

8. Find *x*.

   10°

9. Find *y*.

   105°

10. Find *z*.

   15°

Copyright © by Holt, Rinehart and Winston.
All rights reserved.

LECCIÓN **7-6**

Práctica
Congruencia

Escribe un enunciado de congruencia para cada par de polígonos.

1.

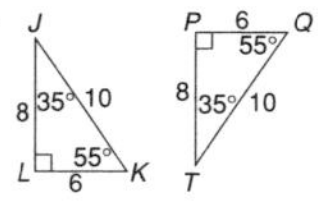

triángulo *JKL* ≅ triángulo *TQP*

2.

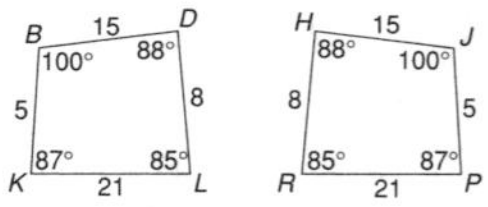

cuadrilátero *BDLK* ≅

cuadrilátero *JHRP*

3.

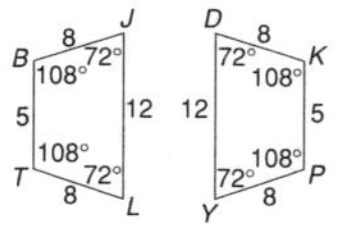

trapecio *BJLT* ≅

trapecio *KDYP*

4.

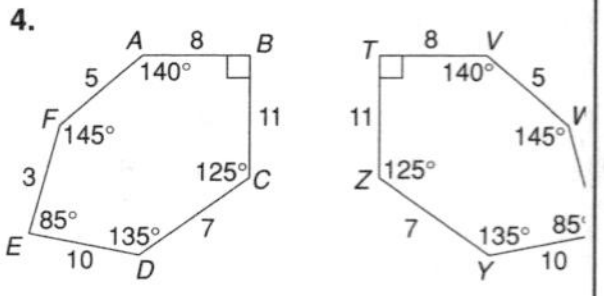

hexágono *ABCDEF* ≅

hexágono *VTZYXW*

En la figura, triángulo *PRT* ≅ triángulo *FJH*.

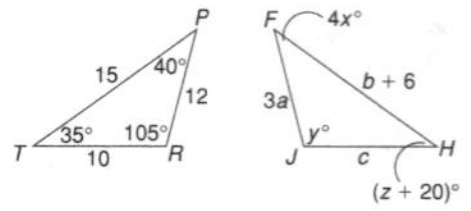

5. Halla *a*.

   4

6. Halla *b*.

   9

7. Halla *c*.

   10

8. Halla *x*.

   10°

9. Halla *y*.

   105°

10. Halla *z*.

   15°

Copyright © by Holt, Rinehart and Winston.
All rights reserved.

Copyright © by Holt, Rinehart and Winston.
All rights reserved.

LESSON 7-7 Practice
Transformations

Identify each as a translation, rotation, reflection, or none of these.

1. 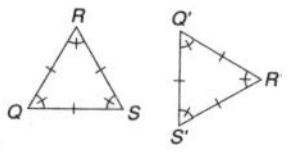

rotation

2. 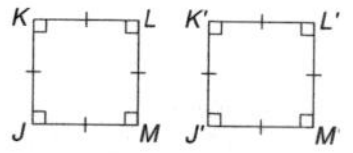

translation

Draw the image of the rectangle *ABCD* with vertices (−2, 1), (−1, 3), and (3, 3), (2, 1) after each transformation.

3. translation 3 units down

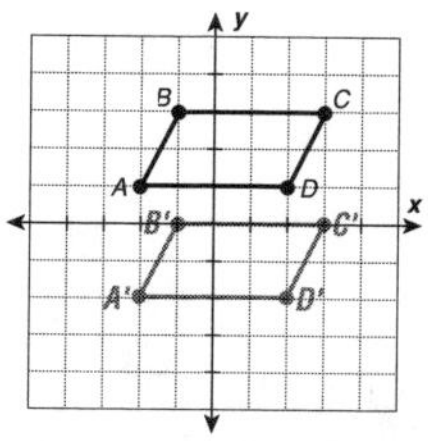

4. 180° rotation around (0, 0)

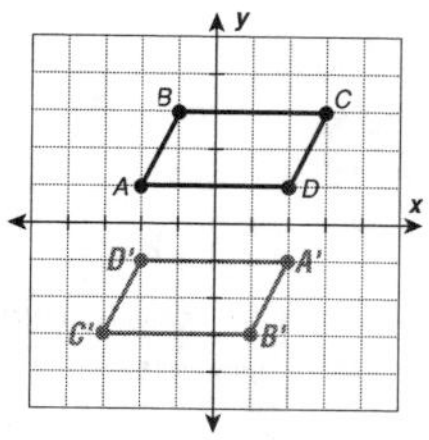

Triangle *ABC* has vertices *A*(−3, 1), *B*(2, 4), and *C*(3, 1). Find the coordinates of the image of each point after each transformation.

5. reflection across the *x*-axis, point *B*

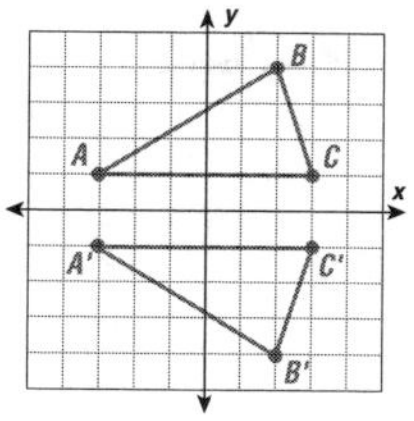

B′(2, −4)

6. translation 6 units down, point *A*

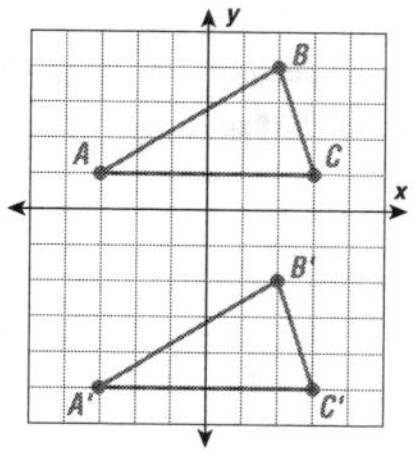

A′(−3, −5)

LECCIÓN 7-7 Práctica
Transformaciones

Identifica si cada ejercicio es una traslación, una rotación, una reflexión o ninguna de las tres cosas.

1. 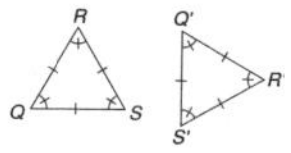

rotación

2. 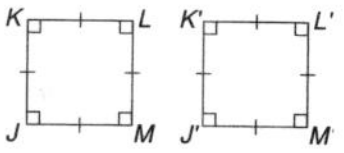

traslación

Dibuja la imagen del rectángulo *ABCD* con vértices (−2, 1), (−1, 3), y (3, 3), (2, 1) después de cada transformación.

3. traslación de 3 unidades hacia abajo

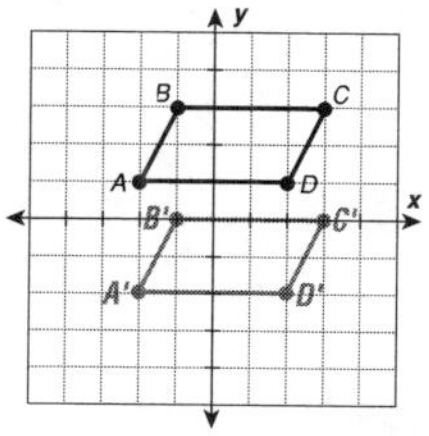

4. rotación de 180° alrededor de (0, 0)

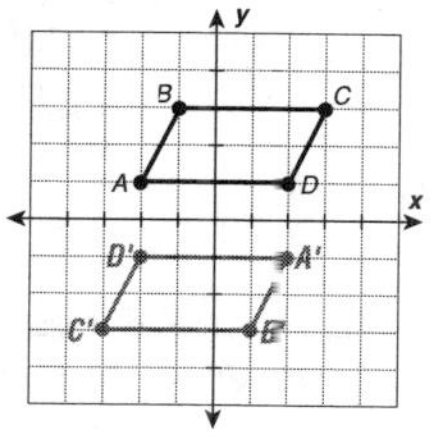

El triángulo *ABC* tiene los vértices *A*(−3, 1), *B*(2, 4), y *C*(3, 1). Halla las coordenadas de la imagen de cada punto después de la transformación.

5. reflexión sobre el eje *x*, punto *B*

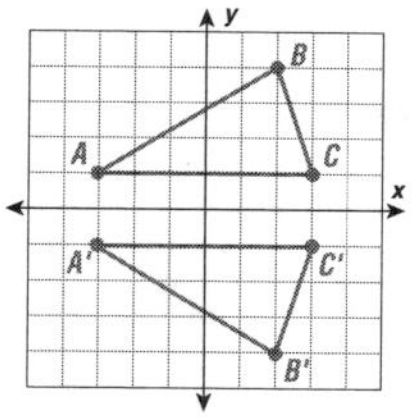

B′(2, −4)

6. traslación de 6 unidades hacia abajo, punto *A*

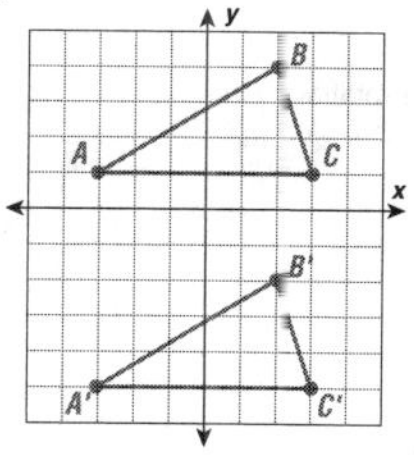

A′(−3, −5)

LESSON 7-8 Practice
Symmetry

Complete each figure. The dashed line is the line of symmetry.

1. 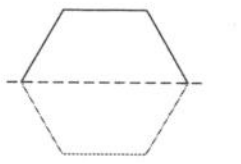

2. 

3. 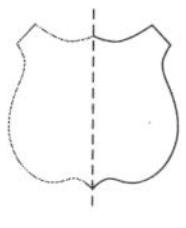

4. 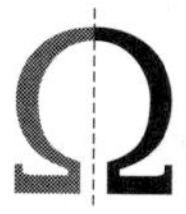

5. 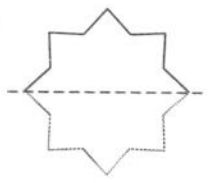

6. 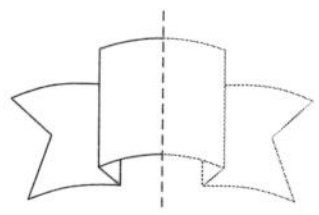

Complete each figure. The point is the center of rotation.

7. 5-fold

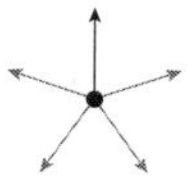

8. 4-fold

9. 2-fold

10. 2-fold

LECCIÓN 7-8 Práctica
Simetría

Completa cada figura. La línea discontinua es el eje de simetría.

1. 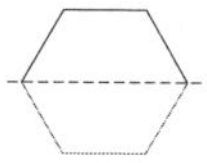

2. 

3. 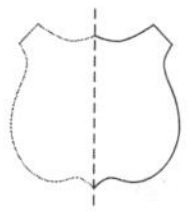

4. 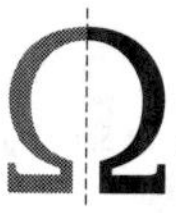

5. 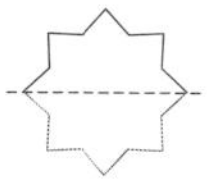

6. 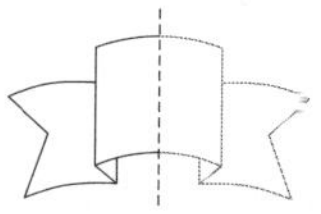

Completa cada figura. El punto es el centro de rotación.

7. de orden 5

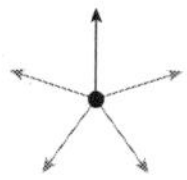

8. de orden 4

9. de orden 2

10. de orden 2

LESSON 7-9

Practice

Tessellations

1. Create a tessellation with quadrilateral *ABCD*.

sample answer:

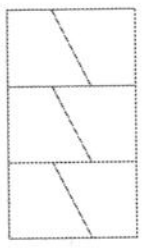

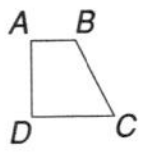

2. Use rotations to create a variation of the tessellation in Exercise 1.

sample answer:

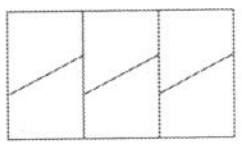

3. Create a tessellation with hexagon ABCDEF.

sample answer:

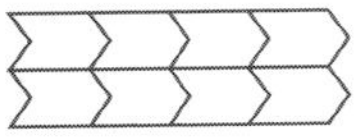

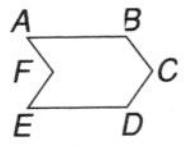

4. Use rotations to create a variation of the tessellation in Exercise 3.

sample answer:

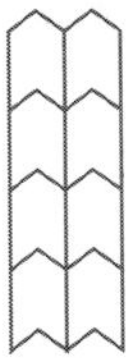

Copyright © by Holt, Rinehart and Winston.
All rights reserved.

LECCIÓN 7-9

Práctica

Teselados

1. Crea un teselado con el cuadrilátero *ABCD*.

ejemplo de respuesta:

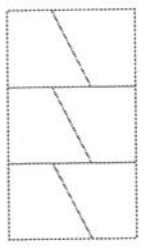

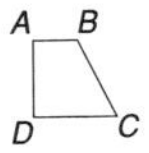

2. Usa rotaciones para crear una variación del teselado del Ejercicio 1.

ejemplo de respuesta:

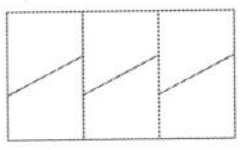

3. Crea un teselado con el hexágono ABCDEF.

ejemplo de respuesta:

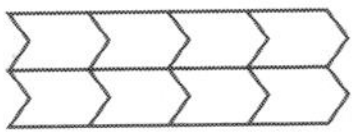

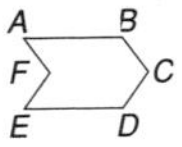

4. Usa rotaciones para crear una variación del teselado del ejercicio 3.

ejemplo de respuesta:

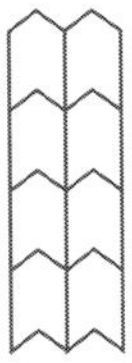

Copyright © by Holt, Rinehart and Winston.
All rights reserved.

LESSON 8-1

Practice

Perimeter and Area of Rectangles and Parallelograms

Find the perimeter of each figure.

1. 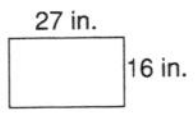

86 in.

2. 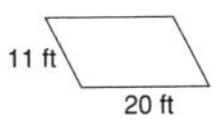

62 ft

3. 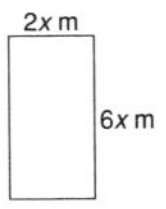

16*x* m

Graph and find the area of each figure with the given vertices.

4. (−3, 4), (3, 4), (3, −4), (−3, −4)

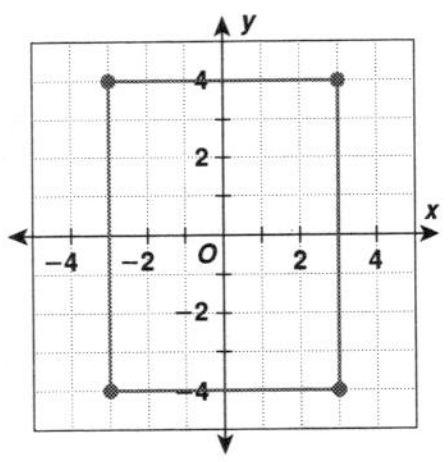

48 units^2

5. (−1, 3), (2, 3), (−1, −4), (−4, −4)

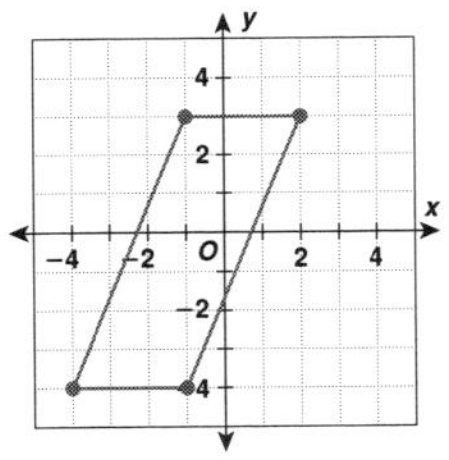

21 units^2

6. Sloppi and Sons Painting Co. charges its customers $1.50 per square foot. How much would Sloppi and Sons charge to paint the rooms of this house if the walls in each room are 9 ft high?

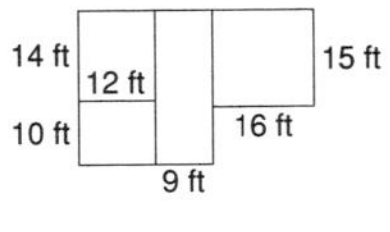

$3024

Copyright © by Holt, Rinehart and Winston.
All rights reserved.

LECCIÓN 8-1

Práctica

Perímetro y área de rectángulos y paralelogramos

Halla el perímetro de cada figura.

1. 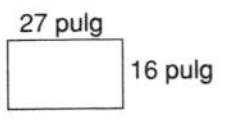

86 pulg

2. 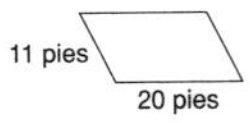

62 pies

3. 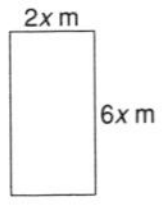

16*x* m

Representa gráficamente y halla el área de cada figura con los vértices dados.

4. (−3, 4), (3, 4), (3, −4), (−3, −4)

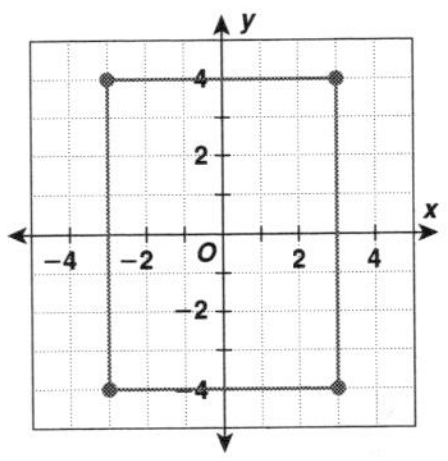

48 unidades^2

5. (−1, 3), (2, 3), (−1, −4), (−4, −4)

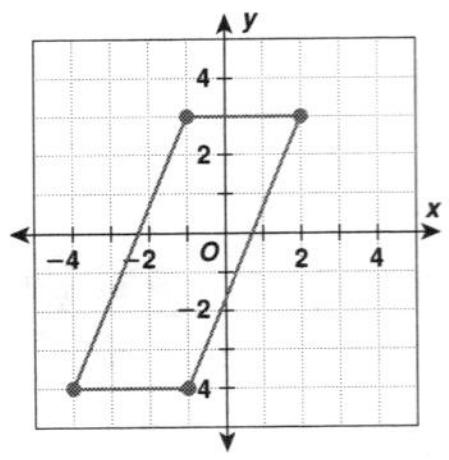

21 unidades^2

6. Pinturas Sloppi e Hijos cobra $1.50 a sus clientes por pie cuadrado. ¿Cuánto cobrará para pintar las habitaciones de esta casa si las paredes de las habitaciones miden 9 pies de altura?

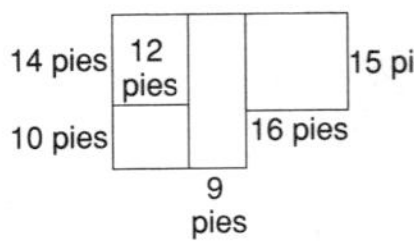

$3024

Copyright © by Holt, Rinehart and Winston.
All rights reserved.

Copyright © by Holt, Rinehart and Winston.
All rights reserved.

LESSON **8-2**
Practice
Perimeter and Area of Triangles and Trapezoids

Find the perimeter of each figure.

1.
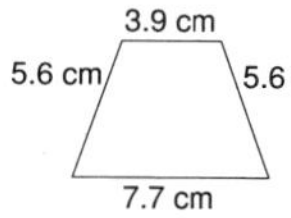

22.8 cm

2.
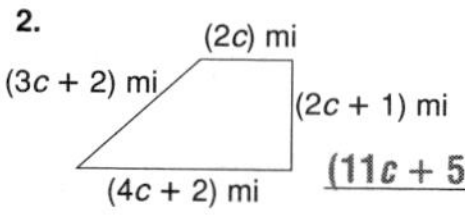

(11c + 5) mi

Find the missing measurement for each figure with the given perimeter.

3. triangle with perimeter 54 units
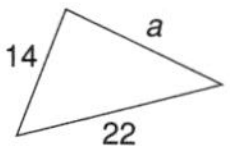

a = 18 units

4. trapezoid with perimeter 34 units
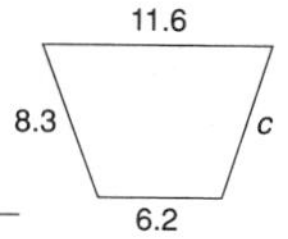

c = 7.9 units

Graph and find the area of each figure with the given vertices.

5. (−1, 3), (4, 3), (4, −4), (−4, −4)
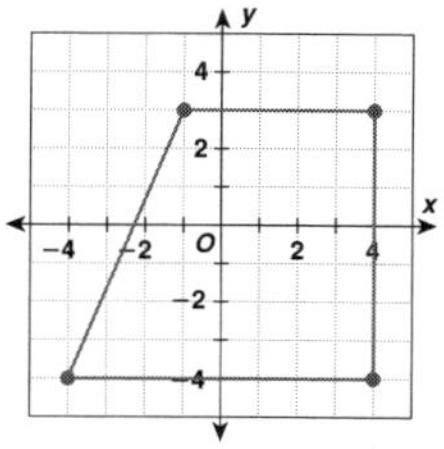

45.5 units^2

6. (−1, 2), (−4, −2), (4, −2)
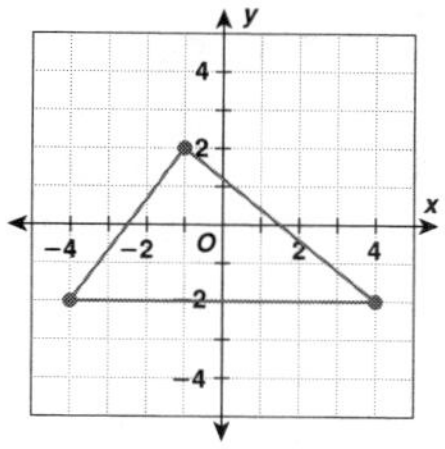

16 units^2

7. The two shortest sides of a pennant shaped like a right triangle measure 10 inches and 24 inches. Hank wants to put colored tape around the edge of the pennant. How many inches of tape does he need?

60 in.

Copyright © by Holt, Rinehart and Winston. All rights reserved.

LECCIÓN **8-2**
Práctica
Perímetro y área de triángulos y trapecios

Halla el perímetro de cada figura.

1.
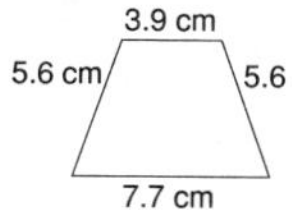

22.8 cm

2.
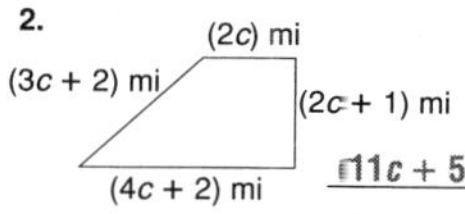

(11c + 5) mi

Halla la medida que falta para cada figura con el perímetro dado.

3. triángulo con un perímetro de 54 unidades
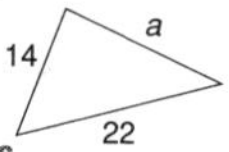

a = 18 unidades

4. trapecio con un perímetro de 34 unidades
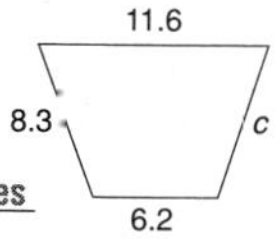

c = 7.9 unidades

Representa gráficamente y halla el área de cada figura con los vértices dados.

5. (−1, 3), (4, 3), (4, −4), (−4, −4)
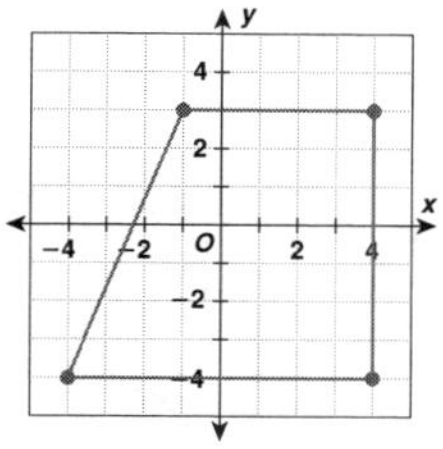

45.5 unidades^2

6. (−1, 2), (−4, −2), (4, −2)
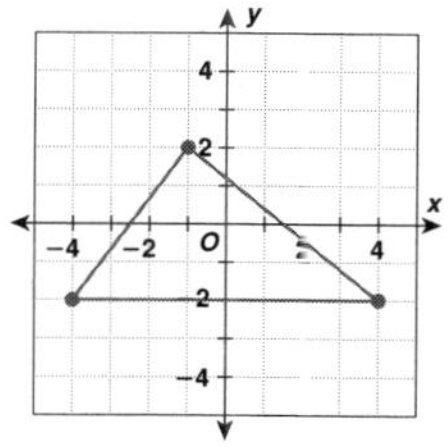

16 unidades^2

7. Los dos lados más cortos de un banderín con forma de triángulo rectángulo miden 10 pulgadas y 24 pulgadas. Hank quiere poner una cinta de color alrededor del borde del banderín. ¿Cuántas pulgadas de cinta necesita?

60 pulg

Copyright © by Holt, Rinehart and Winston. All rights reserved.

LESSON **8-3**
Practice
Circles

Find the circumference of each circle, both in terms of π and to the nearest tenth. Use 3.14 for π.

1. circle with radius 10 in.
20π in. or 62.8 in.

2. circle with diameter 13 cm
13π cm or 40.8 cm

3. circle with diameter 18 m
18π m or 56.5 m

4. circle with radius 15 ft
30π ft or 94.2 ft

5. circle with radius 11.5 in.
23π in. or 72.2 in.

6. circle with diameter 16.4 cm
16.4π cm or 51.5 cm

Find the area of each circle, both in terms of π and to the nearest tenth. Use 3.14 for π.

7. circle with radius 9 in.
81π in^2 or 254.3 in^2

8. circle with diameter 14 cm
49π cm^2 or 153.9 cm^2

9. circle with radius 20 ft
400π ft^2 or 1256 ft^2

10. circle with diameter 17 m
72.3π m^2 or 226.9 m^2

11. circle with diameter 15.4 m
59.3π m^2 or 186.2 m^2

12. circle with radius 22 yd
484π yd^2 or 1519.8 yd^2

13. Graph a circle with center (0, 0) that passes through (0, −3). Find the area and circumference, both in terms of π and to the nearest tenth. Use 3.14 for π.

$A = 9\pi$ units^2 or 28.3 units^2;
$C = 6\pi$ units or 18.8 units

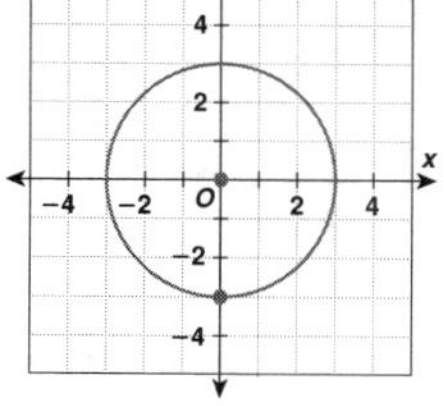

14. A wheel has a radius of $2\frac{1}{3}$ feet. About how far does it travel if it makes 60 complete revolutions? Use $\frac{22}{7}$ for π.

880 ft

Copyright © by Holt, Rinehart and Winston. All rights reserved.

LECCIÓN **8-3**
Práctica
Círculos

Halla la circunferencia de cada círculo en función de π y a la décima más cercana. Usa 3.14 para π.

1. círculo con un radio de 10 pulg
20π pulg ó 62.8 pulg

2. círculo con un diámetro de 13 cm
13π cm ó 40.8 cm

3. círculo con un diámetro de 18 m
18π m ó 56.5 m

4. círculo con un radio de 15 pies
30π pies ó 94.2 pies

5. círculo con un radio de 11.5 pulg
23π pulg ó 72.2 pulg

6. círculo con un diámetro de 16.4 cm
16.4π cm ó 51.5 cm

Halla el área de cada círculo en función de π y a la décima más cercana. Usa 3.14 para π.

7. círculo con un radio de 9 pulg
81π pulg^2 ó 254.3 pulg^2

8. círculo con un diámetro de 14 cm
49π cm^2 ó 153.9 cm^2

9. círculo con un radio de 20 pies
400π pies^2 ó 1256 pies^2

10. círculo con un diámetro de 17 m
72.3π m^2 ó 226.9 m^2

11. círculo con un diámetro de 15.4 m
59.3π m^2 ó 186.2 m^2

12. círculo con un diámetro de 22 yd
484π yd^2 ó 1519.8 yd^2

13. Representa gráficamente un círculo con centro (0, 0) que pasa por el punto (0, −3). Halla el área y la circunferencia en función de π y a la décima más cercana. Usa 3.14 para π.

$A = 9\pi$ unidades^2 ó
28.3 unidades^2;
$C = 6\pi$ unidades ó 18.8 unidades

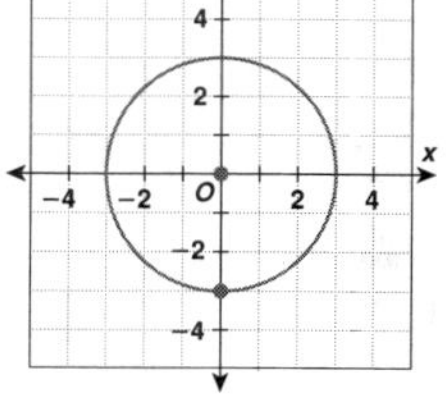

14. Una rueda tiene un radio de $2\frac{1}{3}$ pies. ¿Qué distancia recorre aproximadamente si completa 60 revoluciones? Usa $\frac{22}{7}$ para π.

880 pies

Copyright © by Holt, Rinehart and Winston. All rights reserved.

Copyright © by Holt, Rinehart and Winston. All rights reserved.

LESSON 8-4

Practice
Drawing Three-Dimensional Figures

1. Name the vertices, edges, and faces of the three-dimensional figure shown.

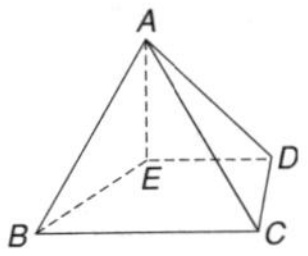

vertices: *A, B, C, D, E*

edges: $\overline{AB}$, $\overline{AC}$, $\overline{AD}$, $\overline{AE}$, $\overline{BC}$, $\overline{CD}$, $\overline{DE}$, $\overline{EB}$

faces: triangles *ABC, ACD, ADE, ABE*, rectangle *BCDE*

2. Draw the figure that has the following top, front, and side views.

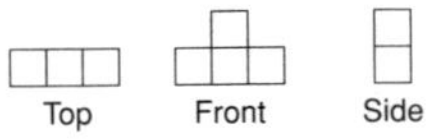

3. Draw the front, top, and side views of the figure.

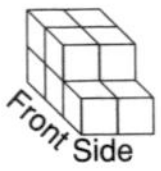

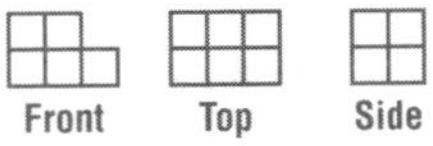

Copyright © by Holt, Rinehart and Winston. All rights reserved.

LECCIÓN 8-4

Práctica
Cómo dibujar figuras tridimensionales

1. Nombra los vértices, aristas y caras de la figura tridimensional que se muestra a continuación.

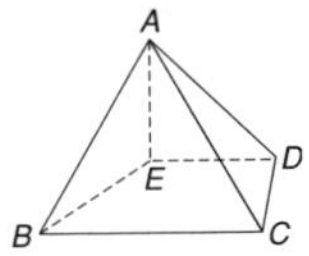

vértices: *A, B, C, D, E*

aristas: $\overline{AB}$, $\overline{AC}$, $\overline{AD}$, $\overline{AE}$, $\overline{BC}$, $\overline{CD}$, $\overline{DE}$, $\overline{EB}$

caras: triángulos *ABC, ACD, ADE, ABE*, rectángulo *BCDE*

2. Dibuja la figura que tenga las siguientes vistas superior, frontal y lateral.

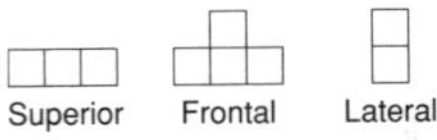

3. Dibuja las vistas frontal, superior y lateral de la figura.

Copyright © by Holt, Rinehart and Winston. All rights reserved.

LESSON 8-5

Practice
Volume of Prisms and Cylinders

Find the volume of each figure to the nearest tenth. Use 3.14 for π.

1.
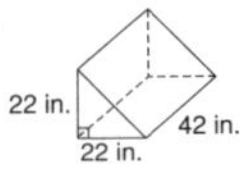

10,164 in^3

2.
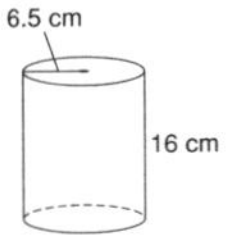

2122.6 cm^3

3.
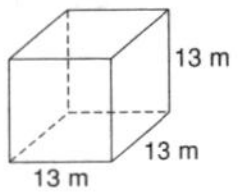

2197 m^3

4.
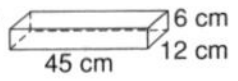

3240 cm^3

5.
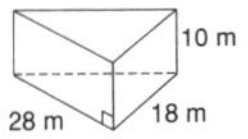

2520 m^3

6.
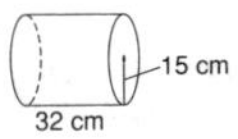

22,608 cm^3

7.
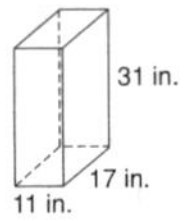

5797 in^3

8.
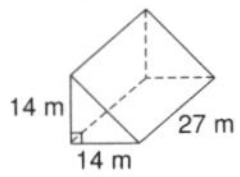

2646 m^3

9.
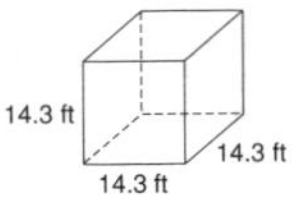

2924.2 ft^3

10. A cylinder has a radius of 6 ft and a height of 25 ft. Explain whether tripling the height will triple the volume of the cylinder. Possible answer: The original cylinder has a volume of 2826 ft^3. If you triple the height the volume is 8478 ft^3, which is triple the original volume.

11. Contemporary American building bricks are rectangular blocks with the standard dimensions of about 5.7 cm by 9.5 cm by 20.3 cm. What is the volume of a brick to the nearest tenth of a unit?

1099.2 cm^3

12. Ian is making candles. His cylindrical mold is 8 in. tall and has a base with a diameter of 3 in. Find the volume of a finished candle to the nearest tenth of a unit.

56.5 in^3

Copyright © by Holt, Rinehart and Winston. All rights reserved.

LECCIÓN 8-5

Práctica
Volumen de prismas y cilindros

Halla el volumen de cada figura a la décima más cercana. Usa 3.14 para π.

1.
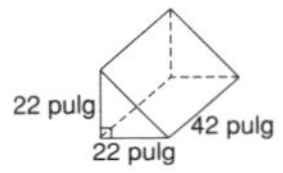

10,164 $pulg^3$

2.
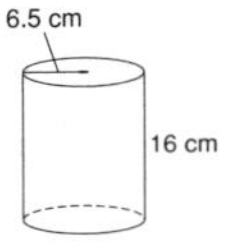

2122.6 cm^3

3.
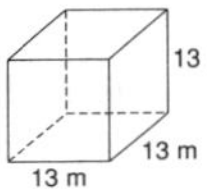

2197 m^3

4.
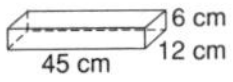

3240 cm^3

5.
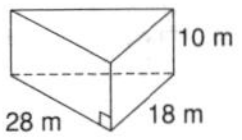

2520 m^3

6.
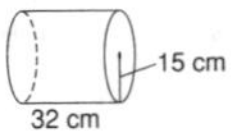

22,608 cm^3

7.
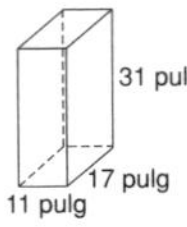

5797 $pulg^3$

8.
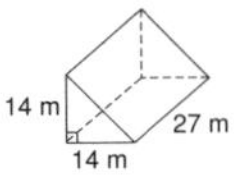

2646 m^3

9.
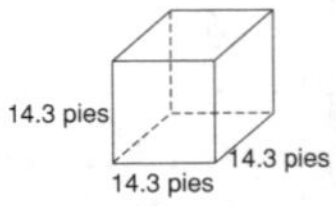

2924.2 $pies^3$

10. Un cilindro tiene un radio de 6 pies y una altura de 25 pies. Explica si triplicar la altura triplicará el volumen del cilindro.

Respuesta posible: El cilindro original tiene un volumen de 2826 $pies^3$. Si se triplica la altura, el volumen es de 8478 $pies^3$, o el triple del volumen original.

11. Los ladrillos de los edificios actuales en Estados Unidos son bloques rectangulares de dimensiones estándar de aproximadamente 5.7 cm por 9.5 cm por 20.3 cm. ¿Cuál es el volumen de un ladrillo a la décima de unidad más cercana?

1099.2 cm^3

12. Ian fabrica velas. El molde cilíndrico mide 8 pulg de altura y el diámetro de la base mide 3 pulg. Halla el volumen de una vela terminada a la décima de unidad más cercana.

56.5 $pulg^3$

Copyright © by Holt, Rinehart and Winston. All rights reserved.

Copyright © by Holt, Rinehart and Winston. All rights reserved.

LESSON **8-6**

Practice

Volume of Pyramids and Cones

Find the volume of each figure to the nearest tenth. Use 3.14 for π.

1. 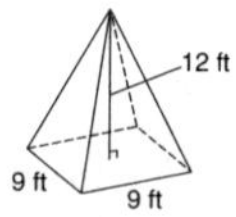

324 ft^3

2. 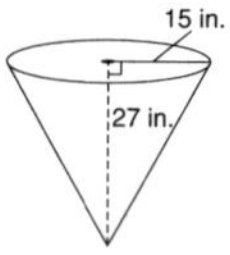

6358.5 in^3

3. 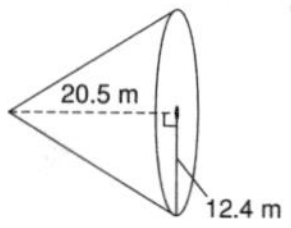

3299.2 m^3

4. 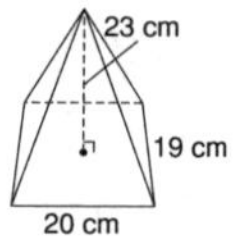

2913.3 cm^3

5. 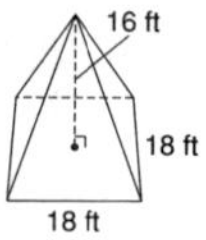

1728 ft^3

6. 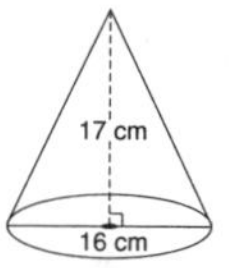

1138.8 cm^3

7. The base of a regular pyramid has an area of 28 in^2. The height of the pyramid is 15 in. Find the volume. **140 in^3**

8. The radius of a cone is 19.4 cm and its height is 24 cm. Find the volume of the cone to the nearest tenth. **9454.2 cm^3**

9. Find the volume of a rectangular pyramid if the height is 13 m and the base sides are 12 m and 15 m. **780 m^3**

10. A funnel has a diameter of 9 in. and is 16 in. deep. Use a calculator to find the volume of the funnel to the nearest hundredth. **339.29 in^3**

11. A square pyramid has a height 18 cm and a base that measures 12 cm on each side. Explain whether tripling the height would triple the volume of the pyramid. **Possible answer:**

The volume of the original pyramid is 864 cm^3. The volume of the new pyramid is 2592 cm^3. Therefore, if the height of the pyramid were tripled, its volume would be tripled.

Copyright © by Holt, Rinehart and Winston. All rights reserved.

LECCIÓN **8-6**

Práctica

Volumen de pirámides y conos

Halla el volumen de cada figura a la décima más cercana. Usa 3.14 para π.

1. 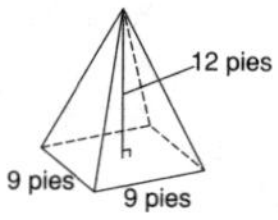

324 pies^3

2. 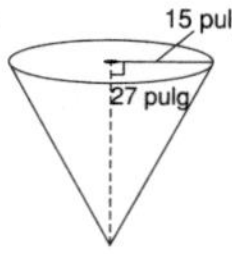

6358.5 pulg^3

3. 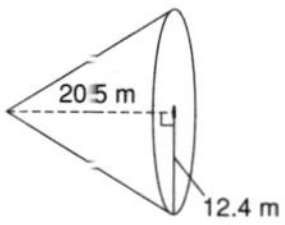

3299.2 m^3

4. 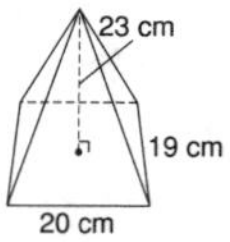

2913.3 cm^3

5. 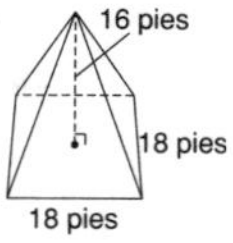

1728 pies^3

6. 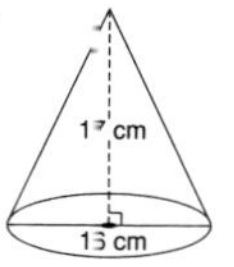

1138.8 cm^3

7. La base de una pirámide regular tiene un área de 28 pulg^2. La altura de la pirámide es de 15 pulg. Halla el volumen. **140 pulg^3**

8. El radio de un cono mide 19.4 cm y su altura es de 24 cm. Halla el volumen del cono a la décima más cercana. **9454.2 cm^3**

9. Halla el volumen de una pirámide rectangular si su altura mide 13 m y los lados de la base miden 12 m y 15 m. **780 m^3**

10. Un embudo tiene un diámetro de 9 pulg y 16 pulg de profundidad. Usa una calculadora para hallar el volumen del embudo a la centésima más cercana. **339.29 pulg^3**

11. Una pirámide cuadrangular tiene una altura de 18 cm y cada lado de la base mide 12 cm. Explica si triplicar la altura triplicaría el volumen de la pirámide.

Respuesta posible: El volumen de la pirámide original es de 864 cm^3. El volumen de la nueva pirámide es de 2592 cm^3. Por lo tanto, si se triplica la altura de la pirámide, se triplicará su volumen.

Copyright © by Holt, Rinehart and Winston. All rights reserved.

LESSON **8-7**

Practice

Surface Area of Prisms and Cylinders

Find the surface area of each figure to the nearest tenth. Use 3.14 for π.

1. 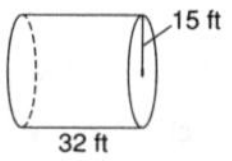

4427.4 ft^2

2. 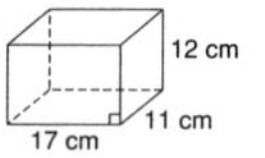

1046 cm^2

3. 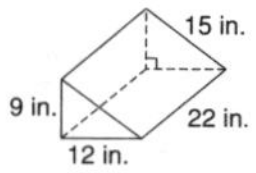

900 in^2

4. 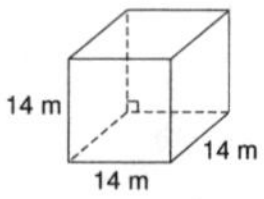

1176 m^2

5. 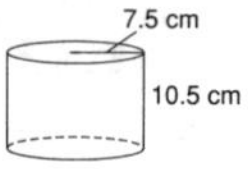

847.8 cm^2

6. 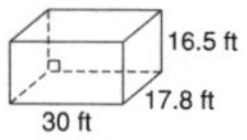

2645.4 ft^2

7. 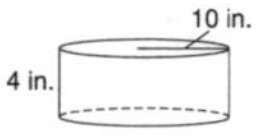

879.2 in^2

8. 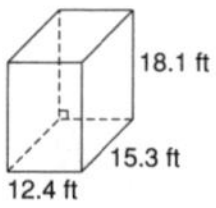

1382.2 ft^2

9. 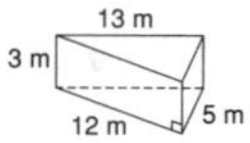

150 m^2

10. Find the surface area to the nearest tenth of a rectangular prism with height 15 m and sides 14 m and 13 m. **1174 m^2**

11. Find the surface area to the nearest tenth of a cylinder 61.7 ft tall that has a diameter of 38 ft. **9629.1 ft^2**

12. Henry wants to paint the ceiling and walls of his living room. One gallon of paint covers 450 ft^2. The room is 24 ft by 18 ft, and the walls are 9 ft high. How many full gallons of paint will Henry need to paint his living room? **3 gal**

13. A rectangular prism is 18 in. by 16 in. by 10 in. Explain the effect, if any, tripling all the dimensions will have on the surface area of the figure. **Possible answer:**

By tripling the dimensions, the surface area becomes 9 times larger, from 1256 in^2 to 11,304 in^2.

Copyright © by Holt, Rinehart and Winston. All rights reserved.

LECCIÓN **8-7**

Práctica

Área total de prismas y cilindros

Halla el área total de cada figura a la décima más cercana. Usa 3.14 para π.

1. 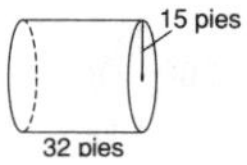

4427.4 pies^2

2. 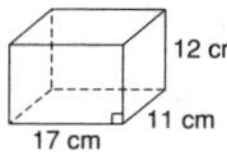

1046 cm^2

3. 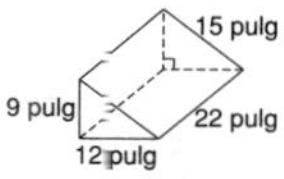

900 pulg^2

4. 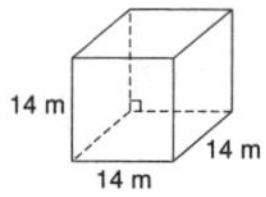

1176 m^2

5. 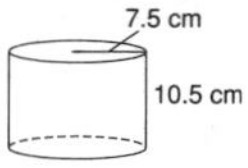

847.8 cm^2

6. 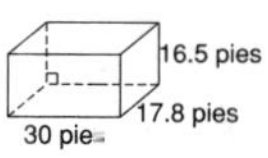

2645.4 pies^2

7. 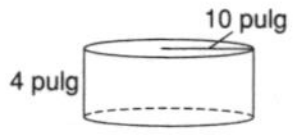

879.2 pulg^2

8. 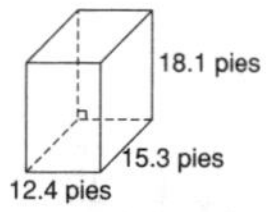

1382.2 pies^2

9. 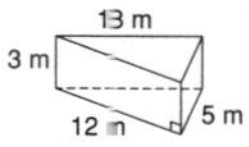

150 m^2

10. Halla el área total de un prisma rectangular con una altura de 15 m y lados de 14 m y 13 m a la décima más cercana. **1174 m^2**

11. Halla el área total de un cilindro de 61.7 pies de altura que tiene un diámetro de 38 pies a la décima más cercana. **9629.1 pies^2**

12. Henry quiere pintar el cielorraso y las paredes de su sala de estar. Un galón de pintura cubre 450 pies^2. La habitación mide 24 pies por 18 pies y las paredes miden 9 pies de altura. ¿Cuántos galones enteros de pintura necesitará Henry para pintar la sala de estar? **3 gal**

13. Un prisma rectangular mide 18 pulg por 16 pulg por 10 pulg. Explica el efecto, si lo hubiera, de triplicar las dimensiones sobre el área total de la figura.

Respuesta posible: Al triplicar las dimensiones, el área total aumenta 9 veces, de 1256 pulg^2 a 11,304 pulg^2.

Copyright © by Holt, Rinehart and Winston. All rights reserved.

Copyright © by Holt, Rinehart and Winston. All rights reserved.

LESSON 8-8

Practice

Surface Area of Pyramids and Cones

Find the surface area of each figure to the nearest tenth. Use 3.14 for π.

1. 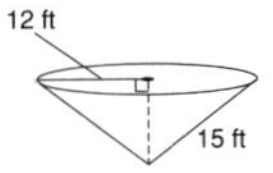

1017.4 ft^2

2. 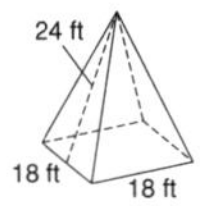

1188 ft^2

3. 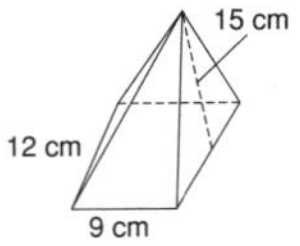

423 cm^2

4. 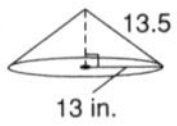

1081.7 in^2

5. 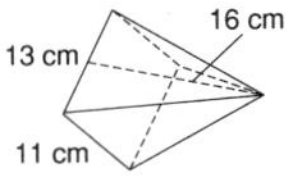

527 cm^2

6. 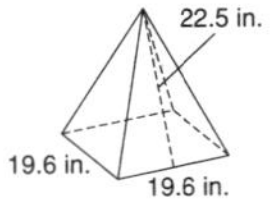

1266.2 in^2

7. 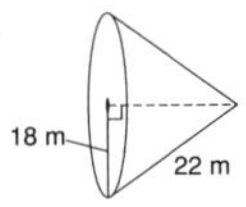

2260.8 m^2

8. 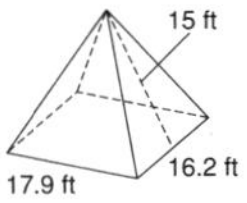

801.5 ft^2

9. 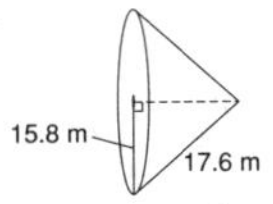

1657.0 m^2

10. Find the surface area of a regular square pyramid with a slant height of 17 m and a base perimeter of 44 m. 495 m^2
11. Find the length of the slant height of a square pyramid if one side of the base is 15 ft and the surface area is 765 ft^2. 18 ft
12. Find the length of the slant height of a cone with a radius of 15 cm and a surface area of 1884 cm^2. 25 cm
13. A cone has a diameter of 12 ft and a slant height of 20 ft. Explain whether tripling both dimensions would triple the surface area. **Possible answer:**

The surface area of the first cone is 489.84 ft^2. The surface area of the cone with the new dimensions is 4408.56 ft^2. It increases the surface area by a factor of 9.

Copyright © by Holt, Rinehart and Winston. All rights reserved.

LECCIÓN 8-8

Práctica

Área total de pirámides y conos

Halla el área total de cada figura a la décima más cercana. Usa 3.14 para π.

1. 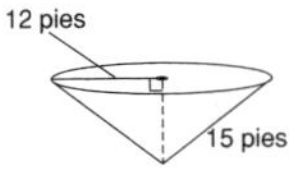

1017.4 pies^2

2. 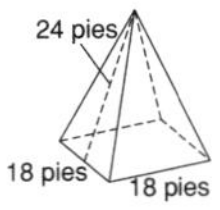

1188 pies^2

3. 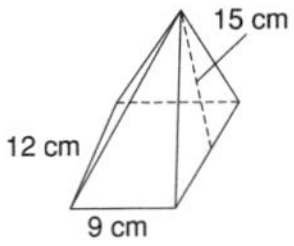

423 cm^2

4. 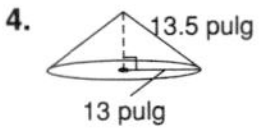

1081.7 pulg^2

5. 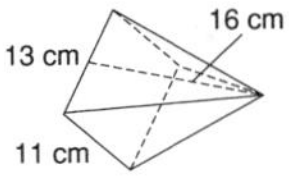

527 cm^2

6. 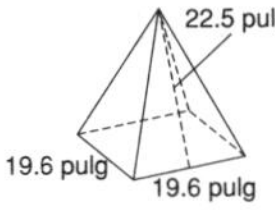

1266.2 pulg^2

7. 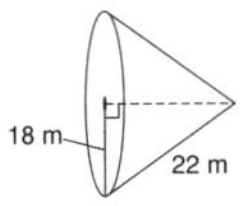

2260.8 m^2

8. 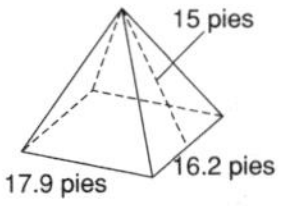

801.5 pies^2

9. 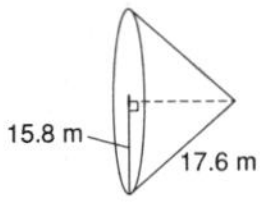

1657.0 m^2

10. Halla el área total de una pirámide cuadrangular regular con una altura inclinada de 17 m y un perímetro de base de 44 m. 495 m^2
11. Halla la longitud de la altura inclinada de una pirámide cuadrangular si un lado de la base mide 15 pies y el área total es de 765 pies^2. 18 pies
12. Halla la longitud de la altura inclinada de un cono con un radio de 15 cm y un área total de 1884 cm^2. 25 cm
13. Un cono tiene un diámetro de 12 pies y una altura inclinada de 20 pies. Explica si triplicar ambas dimensiones triplica el área total.

Respuesta posible: El área total del primer cono es de 489.84 pies^2. El área total del cono con las nuevas dimensiones es de 4408.56 pies^2. El área total aumenta por un factor de 9.

Copyright © by Holt, Rinehart and Winston. All rights reserved.

LESSON 8-9

Practice

Spheres

Find the volume of each sphere, both in terms of π and to the nearest tenth. Use 3.14 for π.

1. $r = 9$ ft — $972\pi \text{ ft}^3 \approx 3052.1 \text{ ft}^3$
2. $r = 21$ m — $12{,}348\pi \text{ m}^3 \approx 38{,}772.7 \text{ m}^3$
3. $d = 30$ cm — $4500\pi \text{ cm}^3 \approx 14{,}130 \text{ cm}^3$
4. $d = 24$ cm — $2304\pi \text{ cm}^3 \approx 7234.6 \text{ cm}^3$
5. $r = 15.4$ in. — $4869.7\pi \text{ in}^3 \approx 15{,}290.8 \text{ in}^3$
6. $r = 16.01$ ft — $5471.6\pi \text{ ft}^3 \approx 17{,}180.8 \text{ ft}^3$

Find the surface area of each sphere, both in terms of π and to the nearest tenth. Use 3.14 for π.

7. 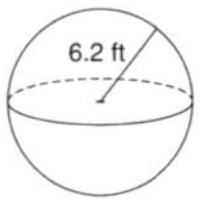

$153.8\pi \text{ ft}^2 \approx 482.8 \text{ ft}^2$

8. 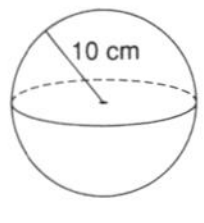

$400\pi \text{ cm}^2 \approx 1256 \text{ cm}^2$

9. 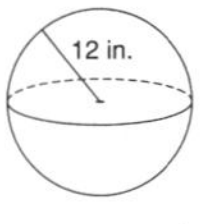

$576\pi \text{ in}^2 \approx 1808.6 \text{ in}^2$

10. 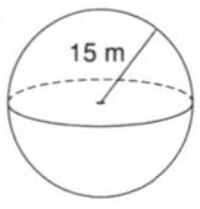

$900\pi \text{ m}^2 \approx 2826 \text{ m}^2$

11. 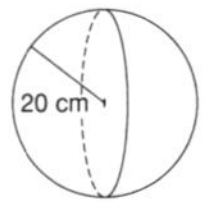

$1600\pi \text{ cm}^2 \approx 5024 \text{ cm}^2$

12. 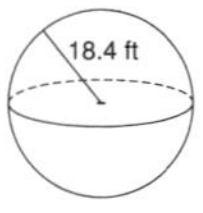

$1354.2\pi \text{ ft}^2 \approx 4252.3 \text{ ft}^2$

13. In the sport of track and field, a field event is the shot put. This is a game in which a heavy ball or shot is thrown or put for distance. The shot itself comes in various sizes, weights and composition. Find the volume and surface area of a shot with diameter 5.5 cm both in terms of π and to the nearest tenth.

volume: $27.7\pi \text{ cm}^3 \approx 87.1 \text{ cm}^3$;
surface area: $30.3\pi \text{ cm}^2 \approx 95.0 \text{ cm}^2$

Copyright © by Holt, Rinehart and Winston. All rights reserved.

LECCIÓN 8-9

Práctica

Esferas

Halla el volumen de cada esfera en función de π y a la décima más cercana. Usa 3.14 para π.

1. $r = 9$ pies — $972\pi \text{ pies}^3 \approx 3052.1 \text{ pies}^3$
2. $r = 21$ m — $12{,}348\pi \text{ m}^3 \approx 38{,}772.7 \text{ m}^3$
3. $d = 30$ cm — $4500\pi \text{ cm}^3 \approx 14{,}130 \text{ cm}^3$
4. $d = 24$ cm — $2304\pi \text{ cm}^3 \approx 7234.6 \text{ cm}^3$
5. $r = 15.4$ pulg — $4869.7\pi \text{ pulg}^3 \approx 15{,}290.8 \text{ pulg}^3$
6. $r = 16.01$ pies — $5471.6\pi \text{ pies}^3 \approx 17{,}180.8 \text{ pies}^3$

Halla el área total de cada esfera en función de π y a la décima más cercana. Usa 3.14 para π.

7. 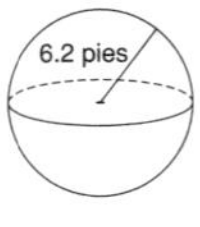

$153.8\pi \text{ pies}^2 \approx 482.8 \text{ pies}^2$

8. 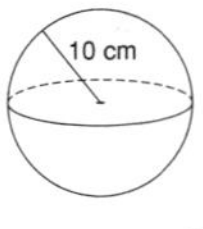

$400\pi \text{ cm}^2 \approx 1256 \text{ cm}^2$

9. 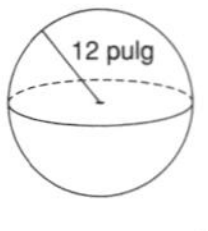

$576\pi \text{ pulg}^2 \approx 1808.6 \text{ pulg}^2$

10. 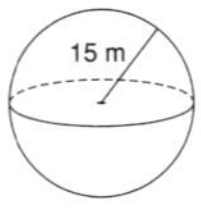

$900\pi \text{ m}^2 \approx 2826 \text{ m}^2$

11. 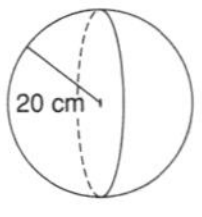

$1600\pi \text{ cm}^2 \approx 5024 \text{ cm}^2$

12. 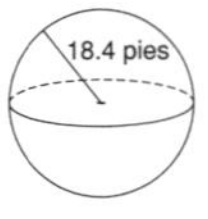

$1354.2\pi \text{ pies}^2 \approx 4252.3 \text{ pies}^2$

13. De las pruebas de atletismo, el lanzamiento de bala es una. En este juego se arroja una pesada esfera o bala lo más lejos posible. La bala viene de distintos tamaños, pesos y materiales. Halla el volumen y el área total de una bala con un diámetro de 5.5 cm en función de π y a la décima más cercana.

volumen: $27.7\pi \text{ cm}^3 \approx 87.1 \text{ cm}^3$; área total: $30.3\pi \text{ cm}^2 \approx 95.0 \text{ cm}^2$

Copyright © by Holt, Rinehart and Winston. All rights reserved.

Copyright © by Holt, Rinehart and Winston. All rights reserved.

LESSON **8-10**

Practice

Scaling Three-Dimensional Figures

A 10 in. cube is built from small cubes, each 2 in. on a side. Compare the following values.

1. The side lengths of the two cubes

   The sides of the 10 in. cube are 5 times as long as the sides of the 2 in. cube.

2. The surface area of the two cubes

   The surface area of the 10 in. cube is 25 times that of the 2 in. cube.

3. The volumes of the two cubes

   The volume of the 10 in. cube is 125 times that of the 2 in. cube.

A 9 cm cube is built from small cubes, each 3 cm on a side. Compare the following values.

4. The side lengths of the two cubes

   The sides of the 9 cm cube are 3 times as long as the sides of the 3 cm cube.

5. The surface area of the two cubes

   The surface area of the 9 cm cube is 9 times that of the 3 cm cube.

6. The volumes of the two cubes

   The volume of the 9 cm cube is 27 times that of the 3 cm cube.

7. The dimensions of a warehouse are 120 ft long, 180 ft wide, and 60 ft high. The scale model used to build the warehouse is 20 in. long. Find the width and height of the model of the warehouse.

   30 in. wide and 10 in. high

8. It takes a machine 40 seconds to fill a cubic box with sides measuring 10 in. How long will it take the same machine to fill a cubic box with sides measuring 15 in.?

   135 seconds

Copyright © by Holt, Rinehart and Winston. All rights reserved.

LECCIÓN **8-10**

Práctica

Hacer dibujos a escala de figuras tridimensionales

Se construye un cubo de 10 pulg con cubos más pequeños, cada uno de 2 pulg de lado. Compara los siguientes valores.

1. El largo de los lados de los dos cubos

   Los lados del cubo de 10 pulg son 5 veces más largos que los lados del cubo de 2 pulg.

2. El área total de los dos cubos

   El área total del cubo de 10 pulg es 25 veces más grande que la del cubo de 2 pulg.

3. El volumen de los dos cubos

   El volumen del cubo de 10 pulg es 125 veces más grande que el del cubo de 2 pulg.

Se construye un cubo de 9 cm con cubos más pequeños, cada uno de 3 cm de lado. Compara los siguientes valores.

4. La longitud de los lados de los dos cubos

   Los lados del cubo de 9 cm son 3 veces más largos que los lados del cubo de 3 cm.

5. El área total de los dos cubos

   El área total del cubo de 9 cm es 9 veces más grande que la del cubo de 3 cm.

6. El volumen de los dos cubos

   El volumen del cubo de 9 cm es 27 veces más grande que el del cubo de 3 cm.

7. Las dimensiones de un depósito son 120 pies de largo, 180 pies de ancho y 60 pies de alto. El modelo a escala usado para construir el depósito mide 20 pulg de largo. Halla el ancho y la altura del modelo a escala.

   30 pulg de ancho y 10 pulg de altura

8. Una máquina tarda 40 segundos en llenar una caja cúbica con lados que miden 10 pulg. ¿Cuánto le llevará a la misma máquina llenar una caja cúbica con lados que miden 15 pulg?

   135 segundos

Copyright © by Holt, Rinehart and Winston. All rights reserved.

LESSON **9-1**

Practice

Samples and Surveys

Identify the sampling method used.

1. People in the security line at the airport are asked to step out of the line for a more detailed search. The people pulled out of the line have not necessarily done anything wrong, and they are not chosen according to any particular rule.

   random

2. At the 1-mile marker of a marathon, a timekeeper shouts out the time elapsed to every 10th runner that passes by. A statistician records the times shouted.

   systematic

3. A geologist visits 10 randomly-selected lakes in the region and collects soil samples in randomly-selected areas along each shoreline.

   stratified

Identify the population and sample. Give a reason the sample could be biased.

4. At a convention of science teachers, various attendees are asked to name their favorite subject in high school.

   population: teachers at the convention

   sample: teachers surveyed

   possible bias: most will say science

5. Donors participating in a blood drive are given a small amount of money for their blood donation. Before they can give blood, each person is surveyed to find out if they are eligible to give blood.

   population: blood donors

   sample: blood donors (entire population)

   possible bias: people may lie to get money

6. Interviewers at the mall are surveying girls with red hair to find out if a correlation exists between personality and red hair.

   population: girls with red hair

   sample: girls surveyed

   possible bias: some girls color their hair

Copyright © by Holt, Rinehart and Winston. All rights reserved.

LECCIÓN **9-1**

Práctica

Muestras y encuestas

Identifica el método de muestreo que se usa.

1. En los controles de seguridad de los aeropuertos se pide a algunas personas que se separen de la fila para realizar una inspección más minuciosa. Estas personas no han hecho necesariamente nada malo y no son elegidos de acuerdo a una regla en particular.

   aleatorio

2. En la marca de una milla de una maratón, un cronometrador anuncia el tiempo que ha transcurrido cada vez que diez corredores pasan a su lado. Un estadístico registra los tiempos que se anuncian.

   sistemático

3. Un geólogo visita 10 lagos en la región elegidos al azar y recolecta muestras de tierra en áreas, también elegidas al azar, a lo largo de las costas de cada lago.

   por estratos

Identifica la población y la muestra. Explica por qué la muestra puede ser no representativa.

4. En una convención de profesores de ciencias, se pide a varios de los presentes que indiquen su materia favorita en la escuela superior.

   población: maestros que asistieron a la convención

   muestra: maestros encuestados

   posible error: la mayoría dirá ciencias

5. Los donantes que participan de una campaña de donación de sangre reciben una modesta cantidad de dinero por su donación. Antes de la extracción, deben responder un cuestionario para ver si están en condiciones de donar sangre.

   población: donantes de sangre

   muestra: donantes de sangre (toda la población)

   posible error: las personas podrían mentir para ganar dinero

6. Varios entrevistadores encuestan a las chicas pelirrojas en un centro comercial para ver si hay alguna correlación entre la personalidad y el cabello rojo.

   población: chicas pelirrojas

   muestra: chicas encuestadas

   posible error: algunas chicas se tiñen el cabello

Copyright © by Holt, Rinehart and Winston. All rights reserved.

Copyright © by Holt, Rinehart and Winston. All rights reserved.

LESSON **9-2**

Practice

Organizing Data

1. Use a line plot to organize the data of the distances students travel to school.

Distances Students Travel to School (mi)

2	8	6	10	5	4	6	8	3	2
11	5	1	3	6	5	7	5	2	4

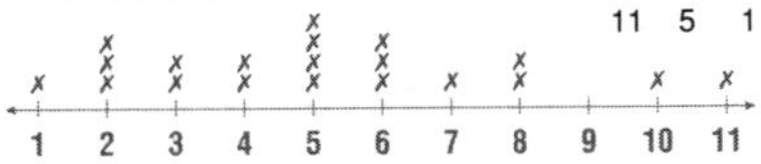

List the data values in the stem-and-leaf plot.

2.

2	0 1 5 7
3	2 2 9
4	5 6 7 9
5	1 3

Key: 5 | 1 = 51

20, 21, 25, 27, 32, 32, 39, 45, 46, 47, 49, 51, 53

3. Use the given data to make a back-to-back stem-and-leaf plot.

NBA Midwest Division 2000–2001 Final Standings

NBA Team	Wins	Losses	NBA Team	Wins	Losses
San Antonio Spurs	58	24	Houston Rockets	45	37
Utah Jazz	53	29	Denver Nuggets	40	42
Dallas Mavericks	53	29	Vancouver Grizzlies	23	59
Minnesota Timberwolves	47	35			

Wins		Losses
3	2	4 9 9
	3	5 7
7 5 0	4	2
8 3 3	5	9

Key: | 5 | 9 represents 59

3 | 5 | represents 53

4. Make a Venn diagram to show how many girls in an eighth-grade class belonged to both a team and a club.

Team	yes	no	yes	no	yes	yes	yes	no	no	yes	no	no
Club	yes	yes	no	yes	yes	no	yes	yes	yes	no	no	yes

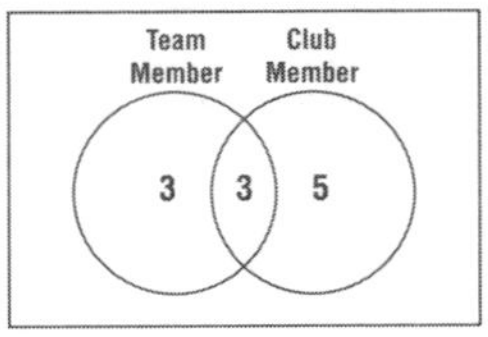

Copyright © by Holt, Rinehart and Winston. All rights reserved.

LECCIÓN **9-2**

Práctica

Cómo organizar datos

1. Usa un diagrama de acumulación para organizar los datos sobre las distancias que recorren los estudiantes para ir a la escuela.

Distancias que recorren los estudiante para ir a la escuela (mi)

2	8	6	10	5	4	6	8	3	2
11	5	1	3	6	5	7	5	2	4

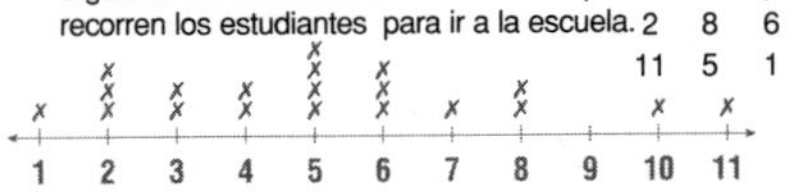

Haz una lista con los valores de los datos del diagrama de tallo y hojas.

2.

2	0 1 5 7
3	2 2 9
4	5 6 7 9
5	1 3

Clave: 5 | 1 = 51

20, 21, 25, 27, 32, 32, 39, 45, 46, 47, 49, 51, 53

3. Usa los datos dados para hacer un diagrama doble de tallo y hojas.

Posición final de la temporada 2000-2001 de la división de la región central de la NBA

Equipo en la NBA	Victorias	Derrotas	Equipo en la NBA	Victorias	Derrotas
San Antonio Spurs	58	24	Houston Rockets	45	37
Utah Jazz	53	29	Denver Nuggets	40	42
Dallas Mavericks	53	29	Vancouver Grizzlies	23	59
Minnesota Timberwolves	47	35			

Victorias		Derrotas
3	2	4 9 9
	3	5 7
7 5 0	4	2
8 3 3	5	9

Clave: | 5 | 9 significa 59

3 | 5 | significa 53

4. Haz un diagrama de Venn para mostrar cuántas estudiantes de octavo grado pertenecían tanto a un equipo como a un club.

Equipo	sí	no	sí	no	sí	sí	sí	no	no	sí	no	no
Club	sí	sí	no	sí	sí	no	sí	sí	sí	no	no	sí

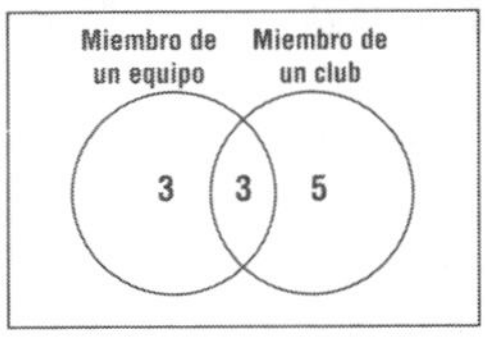

Copyright © by Holt, Rinehart and Winston. All rights reserved.

LESSON **9-3**

Practice

Measures of Central Tendency

Find the mean, median, mode, and range of each data set.

1. 7, 7, 4, 9, 6, 4, 5, 8, 4

mean: 6
median: 6
mode: 4
range: 5

2. 1.2, 5.8, 3.7, 9.7, 5.5, 0.3, 8.1

mean: 4.9
median: 5.5
mode: none
range: 9.4

3. 31, 28, 31, 30, 31, 30, 31, 31, 30, 31, 30, 31

mean: $30.41\overline{6}$
median: 31
mode: 31
range: 3

4. 65, 46, 78, 3, 87, 12, 99, 38, 71, 38

mean: 53.7
median: 55.5
mode: 38
range: 96

Determine and find the most appropriate measure of central tendency or range for each situation. Refer to the table at the right for Exercises 5–7.

Some Major Earthquakes in United States History

Year	Location	Magnitude
1812	Missouri	7.9
1872	California	7.8
1906	California	7.7
1957	Alaska	8.8
1964	Alaska	9.2
1965	Alaska	8.7
1983	Idaho	7.3
1986	Alaska	8.0
1987	Alaska	7.9
1992	California	7.6

5. Which measure best describes the middle of the data?

mean; 8.09

6. Which earthquake magnitude occurred most frequently?

mode; 7.9

7. How spread out are the data?

range; 1.9

8. Nicole purchased gasoline 8 times in the last two months. The prices that she paid per gallon each time were $2.19, $2.14, $2.28, $2.09, $2.01, $1.99, $2.19, and $2.39. Which measure makes the prices appear lowest?

mean; $2.16

Copyright © by Holt, Rinehart and Winston. All rights reserved.

LECCIÓN **9-3**

Práctica

Medidas de tendencia dominante

Halla la media, la mediana, la moda y el rango de cada conjunto de datos.

1. 7, 7, 4, 9, 6, 4, 5, 8, 4

media: 6
mediana: 6
moda: 4
rango: 5

2. 1.2, 5.8, 3.7, 9.7, 5.5, 0.3, 8.1

media: 4.9
mediana: 5.5
moda: ninguna
rango: 9.4

3. 31, 28, 31, 30, 31, 30, 31, 31, 30, 31, 30, 31

media: $30.41\overline{6}$
mediana: 31
moda: 31
rango: 3

4. 65, 46, 78, 3, 87, 12, 99, 38, 71, 38

media: 53.7
mediana: 55.5
moda: 38
rango: 96

Determina y halla la medida de tendencia dominante o el rango más adecuado para cada situación. Consulta la tabla de la derecha para los Ejercicios del 5 al 7.

Algunos de los terremotos de mayor magnitud en la historia de Estados Unidos

Año	Lugar	Magnitud
1812	Missouri	7.9
1872	California	7.8
1906	California	7.7
1957	Alaska	8.8
1964	Alaska	9.2
1965	Alaska	8.7
1983	Idaho	7.3
1986	Alaska	8.0
1987	Alaska	7.9
1992	California	7.6

5. ¿Qué medida describe mejor el valor medio de los datos?

media; 8.09

6. ¿Cuál es la magnitud de terremoto que ocurrió con mayor frecuencia?

moda; 7.9

7. ¿Cuál es la amplitud de los datos?

rango; 1.9

8. Nicole compró combustible 8 veces durante los últimos dos meses. Los precios por galón que pagó cada vez son los siguientes: $2.19, $2.14, $2.28, $2.09, $2.01, $1.99, $2.19 y $2.39. ¿Qué medida hace que los precios parezcan más bajos?

media; $2.16

Copyright © by Holt, Rinehart and Winston. All rights reserved.

Copyright © by Holt, Rinehart and Winston. All rights reserved.

LESSON **9-4**
Practice
Variability

Find the first and third quartiles for each data set.

1. 37, 48, 56, 35, 53, 41, 50

first quartile: 37

third quartile: 53

2. 18, 20, 34, 33, 16, 44, 42, 27

first quartile: 19

third quartile: 38

Use the given data to make a box-and-whisker plot.

3. 55, 46, 70, 36, 43, 45, 52, 61

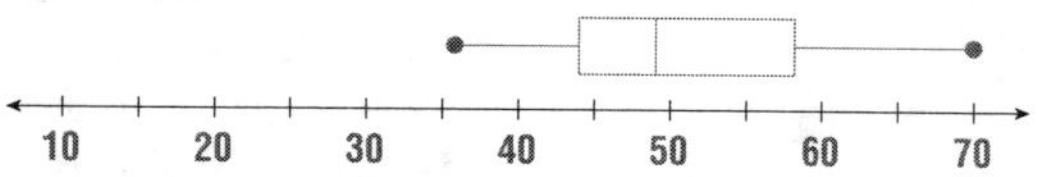

4. 23, 34, 31, 16, 38, 42, 45, 30, 28, 25, 19, 32, 53

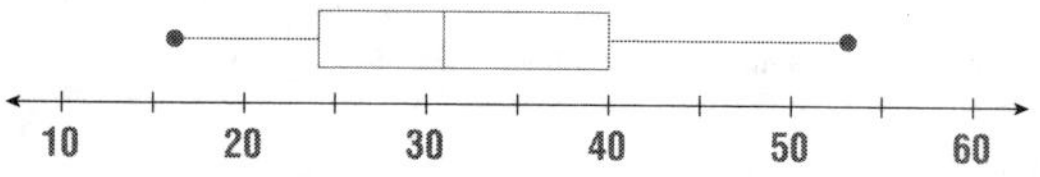

Use the box-and-whisker plots to compare the data sets.

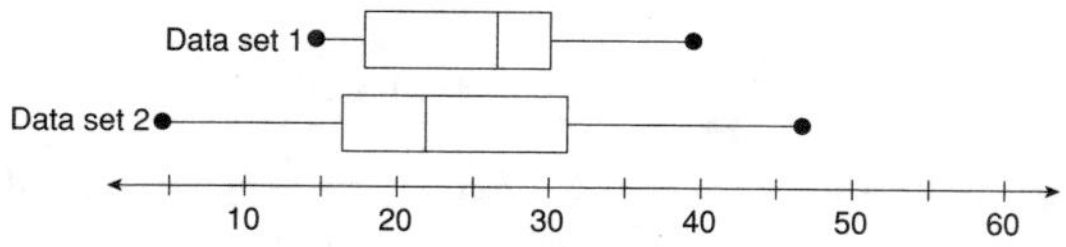

5. Compare the medians and ranges.

The median of data set 1 is greater than the median of data set 2. The range of data set 2 is greater than the range of data set 1.

6. Compare the ranges of the middle half of the data for each set.

The range of the middle half of the data is greater in data set 2.

Copyright © by Holt, Rinehart and Winston. All rights reserved.

LECCIÓN **9-4**
Práctica
Variabilidad

Halla el primer y el tercer cuartil de cada conjunto de datos.

1. 37, 48, 56, 35, 53, 41, 50

primer cuartil: 37

tercer cuartil: 53

2. 18, 20, 34, 33, 16, 44, 42, 27

primer cuartil: 19

tercer cuartil: 38

Usa los siguientes datos para hacer una gráfica de mediana y rango.

3. 55, 46, 70, 36, 43, 45, 52, 61

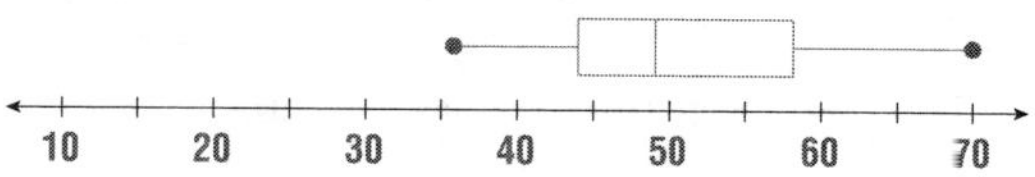

4. 23, 34, 31, 16, 38, 42, 45, 30, 28, 25, 19, 32, 53

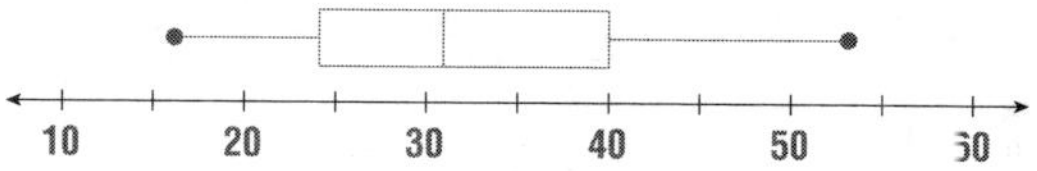

Usa las gráficas de mediana y rango para comparar los conjuntos de datos.

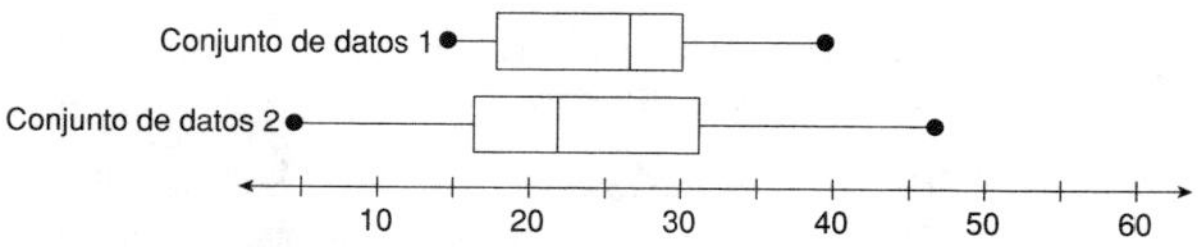

5. Compara las medianas y los rangos.

La mediana del conjunto de datos 1 es mayor que la mediana del conjunto de datos 2. El rango del conjunto de datos 2 es más grande que el rango del conjunto de datos 1.

6. Compara los rangos de la primera mitad de los datos de cada conjunto.

El rango de la primera mitad de los datos es más grande en el conjunto de datos 2.

Copyright © by Holt, Rinehart and Winston. All rights reserved.

LESSON **9-5**
Practice
Displaying Data

1. Make a double-bar graph.

Daily Hours Worked	6	7	8	9	10	11	12
Crew A	4	3	6	1	3	1	2
Crew B	5	5	4	3	2	0	1

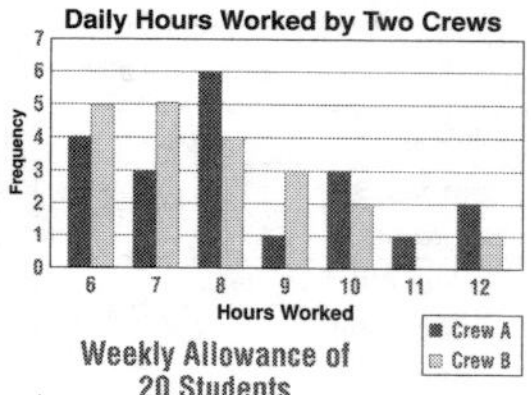

2. Use the data to make a histogram with intervals of 5.

Weekly Allowance of 20 Students			
$5	$15	$2	$10
$12	$12	$10	$15
$10	$5	$6	$4
$8	$7	$20	$7
$5	$4	$5	$9

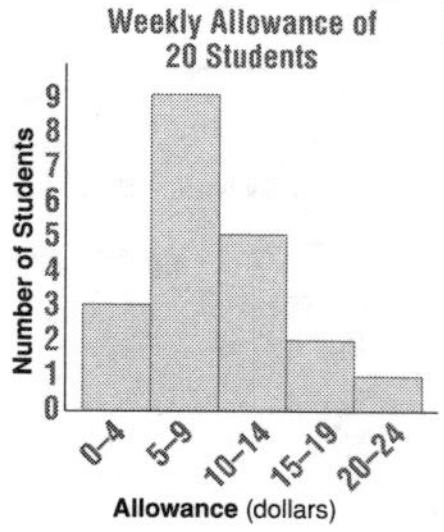

3. Make a double-line graph of the given data. Use the graph to estimate the number of radio stations and cable TV systems in 2002.

Commercial Media in the United States		
Year	Radio Stations	Cable TV Systems
1997	10,207	10,950
1999	10,444	10,700
2001	10,516	9,924
2003	10,605	9,339

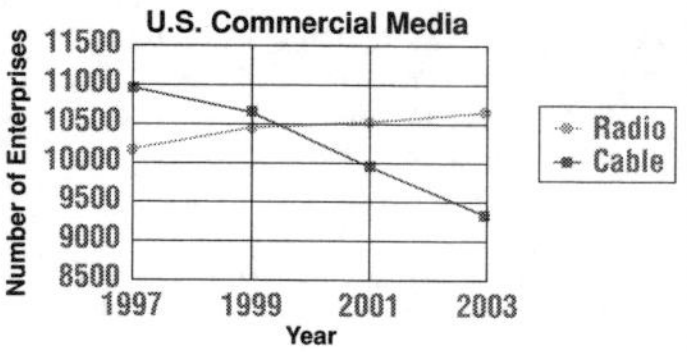

Possible answer: From the graph, I would estimate there were about 10,550 radio stations and about 9,700 cable TV systems in 2002.

Copyright © by Holt, Rinehart and Winston. All rights reserved.

LECCIÓN **9-5**
Práctica
Cómo presentar datos

1. Haz una gráfica de doble barra.

Horas diarias trabajadas	6	7	8	9	10	11	12
Equipo A	4	3	6	1	3	1	2
Equipo B	5	5	4	3	2	0	1

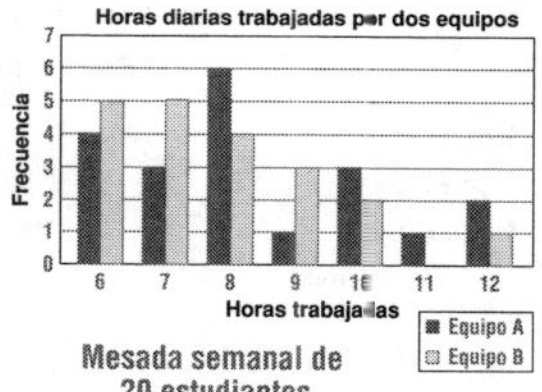

2. Usa los datos para hacer un histograma con intervalos de 5.

Mesada semanal de 20 estudiantes			
$5	$15	$2	$10
$12	$12	$10	$15
$10	$5	$6	$4
$8	$7	$20	$7
$5	$4	$5	$9

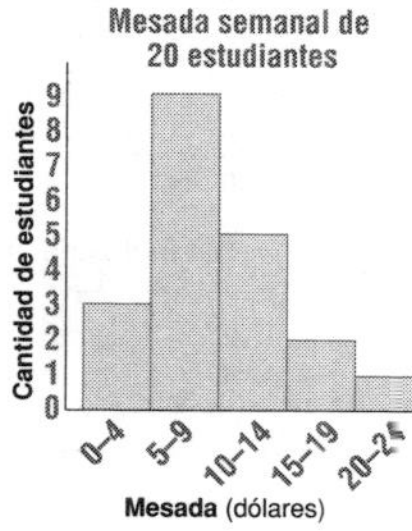

3. Haz una gráfica de doble línea con el siguiente conjunto de datos. Usa la gráfica para estimar la cantidad de estaciones de radio y sistemas de televisión por cable que había en el año 2002.

Medios de difusión comerciales en Estados Unidos		
Año	Estaciones de radio	Sistemas de TV
1997	10,207	10,950
1999	10,444	10,700
2001	10,516	9,924
2003	10,605	9,339

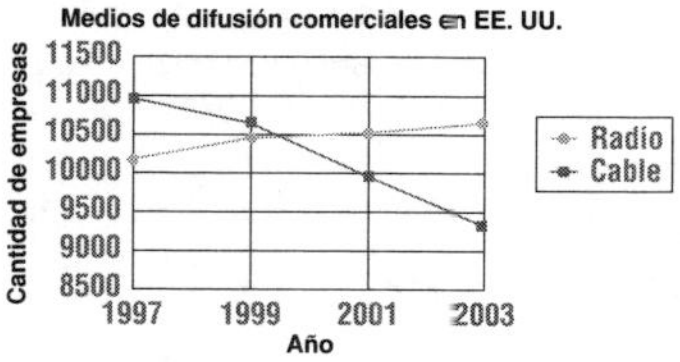

Respuesta posible: A partir de la gráfica, estimo que había alrededor de 10,550 estaciones de radio y 9,700 sistemas de TV por cable en el año 2002.

Copyright © by Holt, Rinehart and Winston. All rights reserved.

Copyright © by Holt, Rinehart and Winston. All rights reserved.

LESSON 9-6

Practice

Misleading Graphs and Statistics

Explain why each graph is misleading.

Possible answers:

1. **On the Road**
Number of Trucks that Travel City Roads

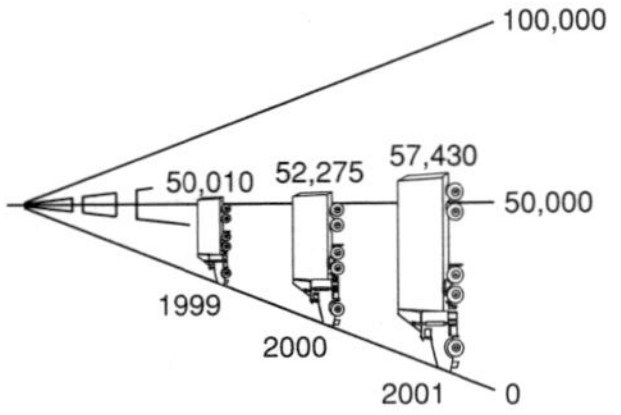

The heights of the truck bars vary greatly because the graph is not equally proportioned. The 2001 bar is about 3 times taller than the 1999 bar. In fact, the data for the 3 years is close together.

2. **Federal Minimun Wage Rates Since 1980**

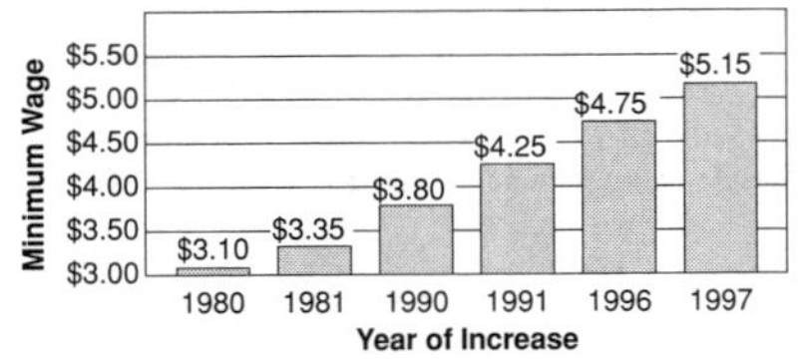

Because the scale does not start at 0, the bar for 1997 is more than 21 times higher than the bar for 1980. In fact, the 1997 minimum wage is only about 1.7 times greater than the 1980 minimum wage.

Explain why the statistic is misleading.

3. A chewing gum company advertises that the flavor of its new chewing gum lasts for an average of 55 minutes based on the following durations reported by customers: 12 min, 33 min, 5 min, 200 min, and 25 min.

Although 55 min is the average, only one person reported that the flavor lasted longer than 33 min. It is not likely that the flavor will last for 55 min.

Copyright © by Holt, Rinehart and Winston.
All rights reserved.

LECCIÓN 9-6

Práctica

Gráficas y estadísticas engañosas

Explica por qué cada una de las siguientes gráficas es engañosa.

Respuesta posible:

1. **En las calles**
Cantidad de camiones que recorren las calles de la ciudad

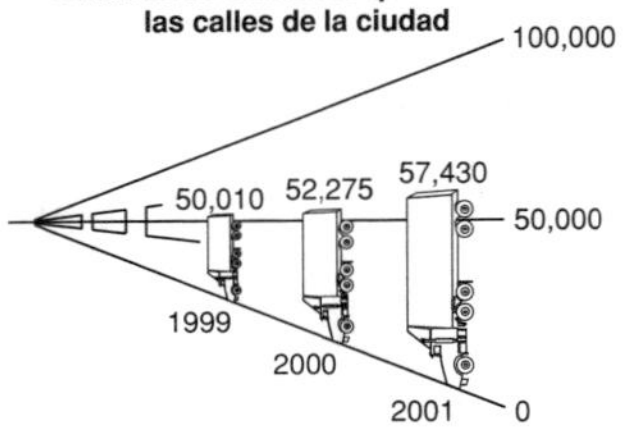

Las alturas de las barras de los camiones varían considerablemente porque la gráfica no está correctamente proporcionada. La barra de 2001 es aproximadamente 3 veces más grande que la barra de 1999. De hecho, los datos de los 3 años son bastante similares.

2. **Salarios mínimos nacionales por hora desde 1980**

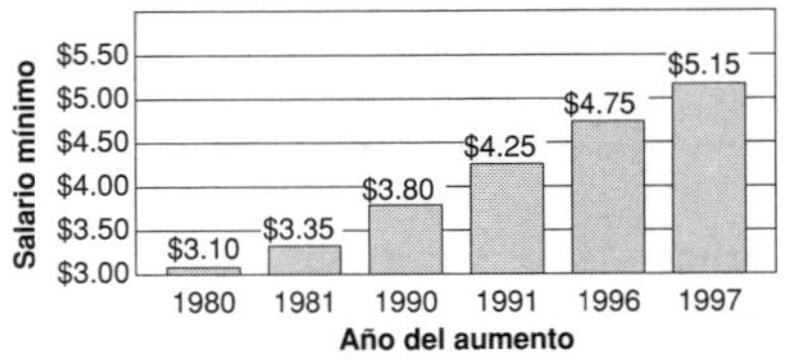

Como la escala no empieza en cero, la barra de 1997 es 21 veces más larga que la barra de 1980. De hecho, el salario mínimo de 1997 es sólo 1.7 veces mayor que el salario mínimo de 1980 aproximadamente.

Explica por qué la siguiente estadística es engañosa.

3. Una empresa de goma de mascar anuncia que el sabor de su nueva goma de mascar dura un promedio de 55 minutos y se basa en los siguientes datos que brindaron los clientes: 12 min, 33 min, 5 min, 200 min y 25 min.

Aunque 55 minutos es el promedio, sólo una persona informó que el sabor duró más de 33 minutos. Es poco probable que el sabor dure 55 minutos.

Copyright © by Holt, Rinehart and Winston.
All rights reserved.

LESSON 9-7

Practice

Scatter Plots

1. Use the given data to make a scatter plot.

Tall Buildings in U.S. Cities

Building	City	Stories	Height (meters)
Sears Tower	Chicago	110	442
Empire State Building	New York	102	381
Bank of America Plaza	Atlanta	55	312
Library Tower	Los Angeles	75	310
Key Tower	Cleveland	57	290
Columbia Seafirst Center	Seattle	76	287
NationsBank Plaza	Dallas	72	281
NationsBank Corporate Center	Charlotte	60	265

Tall Buildings in U.S. Cities

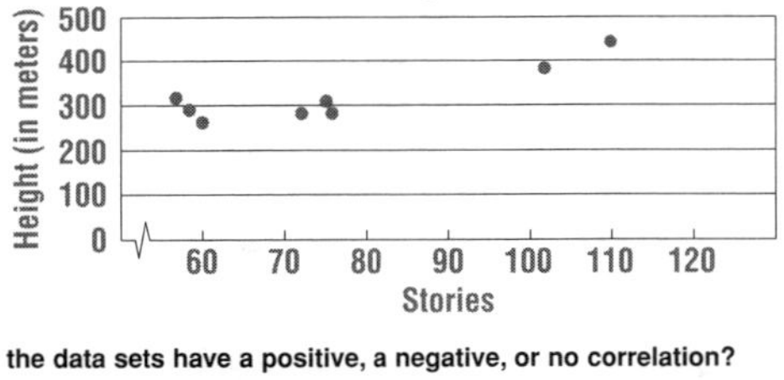

Do the data sets have a positive, a negative, or no correlation?

2. The temperature outside and the number of ice cream cones sold

positive correlation

3. The amount of time spent in the bathtub and the temperature of the bath water

negative correlation

4. Use the data to predict the percent of Americans owning a home in 1955.

Percent of Americans Owning Homes

Year	1950	1960	1970	1980	1990
Percent	55.0%	61.9%	62.9%	64.4%	64.2%

According to the data, about 58.5 % of Americans owned a home in 1955.

Copyright © by Holt, Rinehart and Winston.
All rights reserved.

LECCIÓN 9-7

Práctica

Diagramas de dispersión

1. Usa los siguientes datos para hacer un diagrama de dispersión.

Edificios altos en ciudades de EE.UU.

Edificio	Ciudad	Pisos	Altura (metros)
Torre Sears	Chicago	110	442
Empire State	Nueva York	102	381
Bank of America Plaza	Atlanta	55	312
Library Tower	Los Ángeles	75	310
Tower Key	Cleveland	57	290
Centro Columbia Seafirst	Seattle	76	287
NationsBank Plaza	Dallas	72	281
Centro Corporativo del NationsBank	Charlotte	60	265

Edificios altos en ciudades de EE.UU.

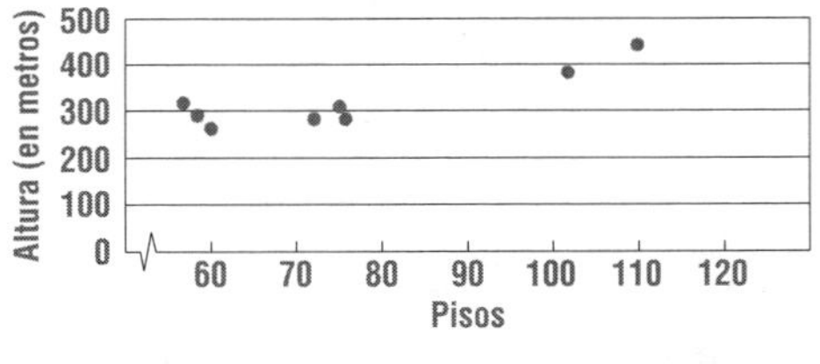

¿Estos conjuntos de datos tienen una correlación positiva, negativa o no hay correlación?

2. La temperatura exterior y la cantidad de barquillos de helado vendidos

correlación positiva

3. El tiempo pasado en la bañera y la temperatura del agua del baño

correlación negativa

4. Usa los datos para predecir el porcentaje de estadounidenses que eran dueños de una casa en 1955.

Porcentaje de estadounidenses propietarios de casas

Año	1950	1960	1970	1980	1990
Porcentaje	55.0%	61.9%	62.9%	64.4%	64.2%

Según los datos, aproximadamente el 58.5 % de los estadounidenses era propietario de una casa en 1955.

Copyright © by Holt, Rinehart and Winston.
All rights reserved.

Copyright © by Holt, Rinehart and Winston.
All rights reserved.

LESSON 9-8
Practice
Choosing the Best Representation of Data

1. Which graph is a better display of the number of students in a class who chose math as their favorite subject?

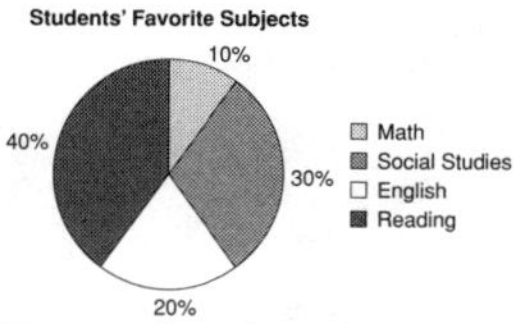

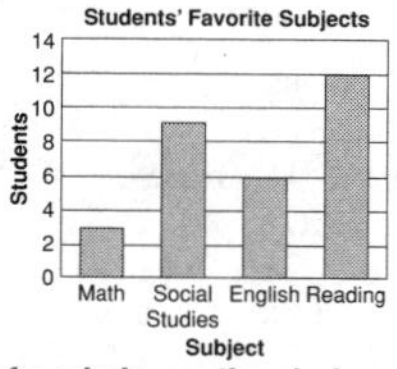

The question asks to find a share, or part of a whole, so the circle graph is a better display.

2. Which graph is a better display of the change in the number of cell telephone subscribers?

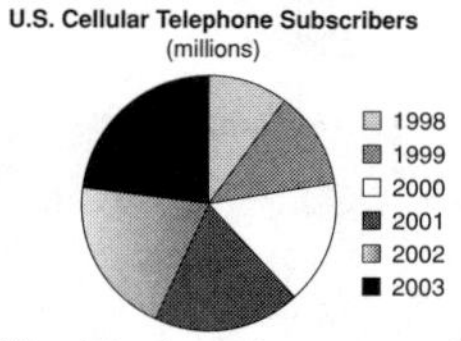

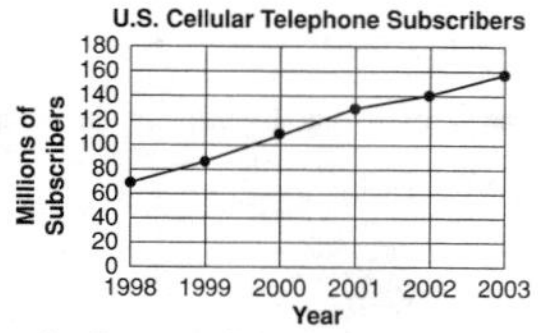

The data show changes over time, so the line graph is the better display.

3. The table shows the heights of players on a school basketball team. Choose an appropriate data display and draw the graph.

Heights of Basketball Players (in.)			
70	64	68	71
61	68	65	73

Possible answer:

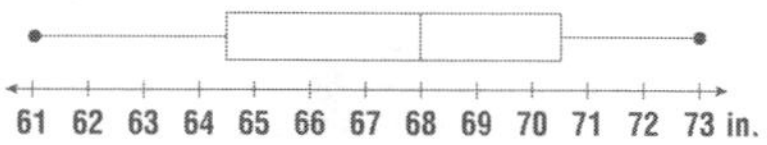

Copyright © by Holt, Rinehart and Winston.
All rights reserved.

LECCIÓN 9-8
Práctica
Cómo elegir la mejor representación de los datos

1. ¿Qué gráfica representa mejor la cantidad de estudiantes de una clase que eligió las matemáticas como su materia favorita?

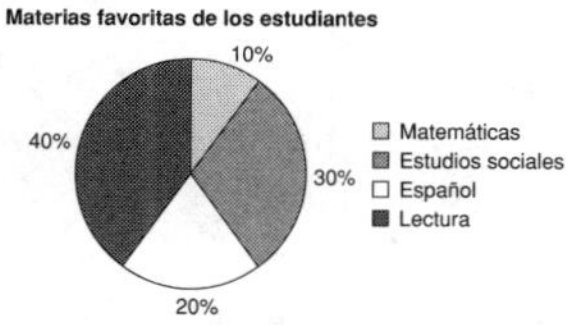

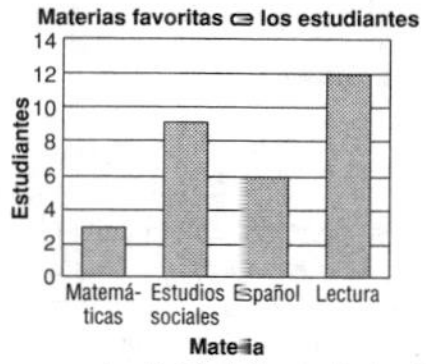

La pregunta requiere mostrar una porción, o una parte del total; por lo tanto, la gráfica circular es la mejor opción.

2. ¿Qué gráfica representa mejor la variación del número de abonados a servicios de telefonía celular?

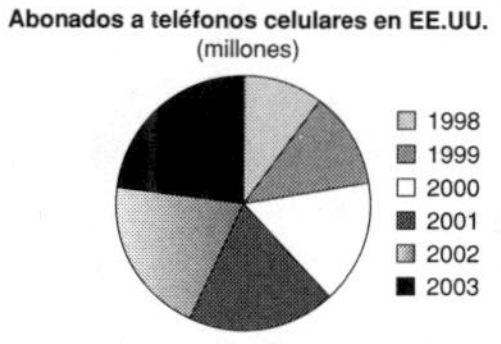

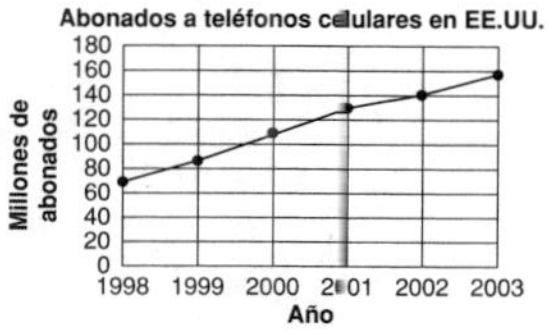

Los datos muestran los cambios a lo largo del tiempo, por lo tanto la gráfica lineal es la mejor opción.

3. En la tabla se muestran las estaturas de los jugadores de un equipo escolar de básquetbol. Elige una forma apropiada de mostrar los datos y dibuja la gráfica.

Estaturas de los jugadores de básquetbol (pulg)			
70	64	68	71
61	68	65	73

Respuesta posible:

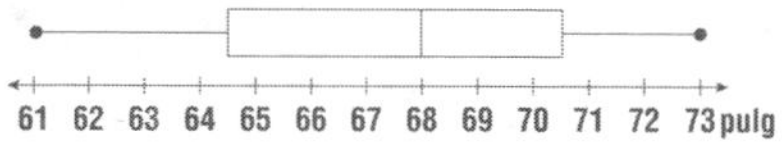

Copyright © by Holt, Rinehart and Winston.
All rights reserved.

LESSON 10-1
Practice
Probability

These are the results of the last math test. The teacher determines that anyone with a grade of more than 70 passed the test. Give the probability for the indicated grade.

Grade	65	70	80	90	100
# of Students	5	3	12	10	2

1. $P(70)$ $\frac{3}{32}$
2. $P(100)$ $\frac{1}{16}$
3. $P(80)$ $\frac{3}{8}$
4. $P(\text{passing})$ $\frac{3}{4}$
5. $P(\text{grade} > 80)$ $\frac{3}{8}$
6. $P(60)$ 0
7. $P(\text{failing})$ $\frac{1}{4}$
8. $P(\text{grade} \leq 80)$ $\frac{5}{8}$

A bowling game consists of rolling a ball and knocking up to 5 pins down. The number of pins knocked down are then counted. The table gives the probability of each outcome.

Number of Pins Down	0	1	2	3	4	5
Probability	0.175	0.189	0.264	0.205	0.132	0.035

9. What is the probability of knocking down all 5 pins?
0.035, or 3.5%
10. What is the probability of knocking down no pins?
0.175, or 17.5%
11. What is the probability of knocking down at most 2 pins?
0.628, or 62.8%
12. What is the probability of knocking down at least 2 pins?
0.636, or 63.6%
13. What is the probability of knocking down more than 3 pins?
0.167, or 16.7%

Copyright © by Holt, Rinehart and Winston.
All rights reserved.

LECCIÓN 10-1
Práctica
Probabilidad

Éstos son los resultados del último examen de matemáticas. El maestro determinó que quienes obtuvieran más de 70 puntos aprobarían el examen. Indica la probabilidad para la puntuación dada.

Puntuación	65	70	80	90	100
Cantidad de estudiantes	5	3	12	10	2

1. $P(70)$ $\frac{3}{32}$
2. $P(100)$ $\frac{1}{16}$
3. $P(80)$ $\frac{3}{8}$
4. $P(\text{aprobado})$ $\frac{3}{4}$
5. $P(\text{puntuación} > 80)$ $\frac{3}{8}$
6. $P(60)$ 0
7. $P(\text{no aprobado})$ $\frac{1}{4}$
8. $P(\text{puntuación} \leq 80)$ $\frac{5}{8}$

Un juego de boliche consiste en arrojar una bola y derribar hasta 5 pinos. Se cuenta la cantidad de pinos derribados. En la tabla se indica la probabilidad de cada resultado.

Cantidad de pinos derribados	0	1	2	3	4	5
Probabilidad	0.175	0.189	0.264	0.205	0.132	0.035

9. ¿Cuál es la probabilidad de derribar los 5 pinos?
0.035, ó 3.5%
10. ¿Cuál es la probabilidad de no derribar ningún pino?
0.175, ó 17.5%
11. ¿Cuál es la probabilidad de derribar como máximo 2 pinos?
0.628, ó 62.8%
12. ¿Cuál es la probabilidad de derribar al menos 2 pinos?
0.636, ó 63.6%
13. ¿Cuál es la probabilidad de derribar más de tres pinos?
0.167, ó 16.7%

Copyright © by Holt, Rinehart and Winston.
All rights reserved.

Copyright © by Holt, Rinehart and Winston.
All rights reserved.

LESSON **10-2**

Practice
Experimental Probability

1. A number cube was thrown 150 times. The results are shown in the table below. Estimate the probability for each outcome.

Outcome	1	2	3	4	5	6
Frequency	33	21	15	36	27	18
Probability	22%	14%	10%	24%	18%	12%

A movie theater sells popcorn in small, medium, large and jumbo sizes. The customers of the first show purchase 4 small, 20 medium, 40 large, and 16 jumbo containers of popcorn. Estimate the probability of the purchase of each of the different size containers of popcorn.

2. *P*(small container) $\frac{1}{20}$ or 5%

3. *P*(medium container) $\frac{1}{4}$ or 25%

4. *P*(large container) $\frac{1}{2}$ or 50%

5. *P*(jumbo container) $\frac{1}{5}$ or 20%

Janessa polled 154 students about their favorite winter sport.

Outcome	Frequency
Skiing	46
Sledding	21
Snowboarding	64
Ice Skating	14
Other	9

6. Use the table to compare the probability that a student chose snowboarding to the probability that a student chose skiing.

≈0.415; ≈0.299; more likely

7. Use the table to compare the probability that a student chose ice skating to the probability that a student chose sledding.

≈0.091; ≈0.136; less likely

8. The class president made 75 copies of the flyer advertising the school play. It was found that 8 of the copies were defective. Estimate the probability that a flyer will be printed properly. ≈0.893

Copyright © by Holt, Rinehart and Winston. All rights reserved.

LECCIÓN **10-2**

Práctica
Probabilidad experimental

1. Se lanzó un dado 150 veces. Los resultados se muestran en la siguiente tabla. Estima la probabilidad para cada resultado.

Resultado	1	2	3	4	5	6
Frecuencia	33	21	15	36	27	18
Probabilidad	22%	14%	10%	24%	18%	12%

Una sala de cine vende palomitas de maíz en paquetes pequeños, medianos, grandes y gigantes. Los clientes de la primera exhibición compran 4 paquetes pequeños, 20 medianos, 40 grandes y 16 gigantes. Estima la probabilidad de compra de cada uno de los distintos tamaños de paquetes de palomitas.

2. *P*(paquete pequeño) $\frac{1}{20}$ ó 5%

3. *P*(paquete mediano) $\frac{1}{4}$ ó 25%

4. *P*(paquete grande) $\frac{1}{2}$ ó 50%

5. *P*(paquete gigante) $\frac{1}{5}$ ó 20%

Janessa encuestó a 154 estudiantes sobre su deporte invernal favorito.

Resultado	Frecuencia
Esquí	46
Trineo	21
Snowboard	64
Patinaje sobre hielo	14
Otros	9

6. Usa los datos de la tabla para comparar la probabilidad de que un estudiante elija snowboard con la probabilidad de que elija esquí.

≈0.415; ≈0.299; más probable

7. Usa los datos de la tabla para comparar la probabilidad de que un estudiante elija patinar sobre hielo con la probabilidad de que elija montar un trineo.

≈0.091; ≈0.136; menos probable

8. El presidente de la clase hizo 75 copias de un volante para anunciar la obra teatral de la escuela. Se descubrió que había 8 copias defectuosas. Estima la probabilidad de que un volante esté bien impreso. ≈0.893

Copyright © by Holt, Rinehart and Winston. All rights reserved.

LESSON **10-3**

Practice
Use a Simulation

Use the table of random numbers for the problems below.

8125	4764	7693	3675	1642	7988	7048	9135	3138	3256
9566	4413	7215	7992	4320	7438	3805	5473	8847	2397
7336	5393	8623	8570	5095	5685	6695	3570	3605	4656
6470	6065	8239	2953	5942	6496	8899	0701	5368	2106
5210	2570	8137	3587	3578	6657	6636	7188	5717	1770
4329	4110	2655	8258	9928	3873	5609	3695	7091	0368
5315	2654	0484	4601	4336	6624	5403	5870	8545	3905
2361	9097	3753	2498	0544	0923	6099	1737	4025	1221
2677	7741	5342	9844	3722	5120	8742	1382	2842	7386
3292	5084	1130	2747	0664	9718	6072	9432	7008	2024

Mr. Domino gave the same math test to all three of his math classes. In the first two classes, 80% of the students passed the test. If the third class has 20 students, estimate the number of students who will pass the test.

1. Using the first row as the first trial, count the successful outcomes and name the unsuccessful outcomes.

16 out of 20 successful; 81, 93, 88, 91

2. Count and name the successful outcomes in the second row as the second trial.

16 out of 20 successful; 95, 92, 88, 97

Determine the successful outcomes in the remaining rows of the random number table.

3. third row 14
4. fourth row 16
5. fifth row 17
6. sixth row 16
7. seventh row 18
8. eighth row 16
9. ninth row 16
10. tenth row 16

11. Based on the simulation, estimate the probability that 80% of the class will pass the math test. 90%

Copyright © by Holt, Rinehart and Winston. All rights reserved.

LECCIÓN **10-3**

Práctica
Usar una simulación

Usa la tabla de números aleatorios para resolver los siguientes problemas.

8125	4764	7693	3675	1642	7988	7048	9135	3138	3256
9566	4413	7215	7992	4320	7438	3805	5473	8847	2397
7336	5393	8623	8570	5095	5685	6695	3570	3605	4656
6470	6065	8239	2953	5942	6496	8899	0701	5368	2106
5210	2570	8137	3587	3578	6657	6636	7188	5717	1770
4329	4110	2655	8258	9928	3873	5609	3695	7091	0368
5315	2654	0484	4601	4336	6624	5403	5870	8545	3905
2361	9097	3753	2498	0544	0923	6099	1737	4025	1221
2677	7741	5342	9844	3722	5120	8742	1382	2842	7386
3292	5084	1130	2747	0664	9718	6072	9432	7008	2024

El señor Domino le entregó el mismo examen a sus tres clases de matemáticas. En las primeras dos clases, 80% de los estudiantes aprobaron el examen. Si en la tercera clase hay 20 estudiantes, estima cuántos van a aprobar el examen.

1. Usando la primera hilera como la primera prueba, cuenta los resultados exitosos e indica los resultados que no tuvieron éxito.

16 de 20 exitosos; 81, 93, 88, 91

2. Cuenta e indica los resultados exitosos en la segunda hilera como segunda prueba.

16 de 20 exitosos; 95, 92, 88, 97

Determina los resultados exitosos en las hileras restantes de la tabla de números aleatorios.

3. tercera hilera 14
4. cuarta hilera 16
5. quinta hilera 17
6. sexta hilera 16
7. séptima hilera 18
8. octava hilera 16
9. novena hilera 16
10. décima hilera 16

11. Basándote en la simulación, estima la probabilidad de que el 80% de la clase apruebe el examen de matemáticas. 90%

Copyright © by Holt, Rinehart and Winston. All rights reserved.

Copyright © by Holt, Rinehart and Winston. All rights reserved.

LESSON **10-4**
Practice
Theoretical Probability

An experiment consists of rolling one fair number cube. Find the probability of each event.

1. $P(3)$ — $\frac{1}{6}$
2. $P(7)$ — 0
3. $P(1 \text{ or } 4)$ — $\frac{1}{3}$
4. $P(\text{not } 5)$ — $\frac{5}{6}$
5. $P(< 5)$ — $\frac{2}{3}$
6. $P(> 4)$ — $\frac{1}{3}$
7. $P(2 \text{ or odd})$ — $\frac{2}{3}$
8. $P(\leq 3)$ — $\frac{1}{2}$

An experiment consists of rolling two fair number cubes. Find the probability of each event.

9. $P(\text{total shown} = 3)$ — $\frac{1}{18}$
10. $P(\text{total shown} = 7)$ — $\frac{1}{6}$
11. $P(\text{total shown} = 9)$ — $\frac{1}{9}$
12. $P(\text{total shown} = 2)$ — $\frac{1}{36}$
13. $P(\text{total shown} = 4)$ — $\frac{1}{12}$
14. $P(\text{total shown} = 13)$ — 0
15. $P(\text{total shown} > 8)$ — $\frac{5}{18}$
16. $P(\text{total shown} \leq 12)$ — 1
17. $P(\text{total shown} < 7)$ — $\frac{5}{12}$

18. A bag contains 9 pennies, 8 nickels, and 5 dimes. How many quarters should be added to the bag so the probability of drawing a dime is $\frac{1}{6}$? — 8 quarters

19. In a game two fair number cubes are rolled. To make the first move, you need to roll a total of 6, 7, or 8. What is the probability that you will be able to make the first move? — $\frac{4}{9}$

Copyright © by Holt, Rinehart and Winston. All rights reserved.

LECCIÓN **10-4**
Práctica
Probabilidad teórica

Un experimento consiste en lanzar un dado. Halla la probabilidad de que ocurra cada suceso.

1. $P(3)$ — $\frac{1}{6}$
2. $P(7)$ — 0
3. $P(1 \text{ ó } 4)$ — $\frac{1}{3}$
4. $P(\text{no } 5)$ — $\frac{5}{6}$
5. $P(< 5)$ — $\frac{2}{3}$
6. $P(> 4)$ — $\frac{1}{3}$
7. $P(2 \text{ o impar})$ — $\frac{2}{3}$
8. $P(\leq 3)$ — $\frac{1}{2}$

Un experimento consiste en lanzar dos dados. Halla la probabilidad de que ocurra cada suceso.

9. $P(\text{total} = 3)$ — $\frac{1}{18}$
10. $P(\text{total} = 7)$ — $\frac{1}{6}$
11. $P(\text{total} = 9)$ — $\frac{1}{9}$
12. $P(\text{total} = 2)$ — $\frac{1}{36}$
13. $P(\text{total} = 4)$ — $\frac{1}{12}$
14. $P(\text{total} = 13)$ — 0
15. $P(\text{total} > 8)$ — $\frac{5}{18}$
16. $P(\text{total} \leq 12)$ — 1
17. $P(\text{total} < 7)$ — $\frac{5}{12}$

18. En una bolsa hay 9 monedas de 1 centavo, 8 monedas de 5 centavos y 5 monedas de 10 centavos. ¿Cuántas monedas de 25 centavos habría que agregar a la bolsa para que la probabilidad de sacar una moneda de 10 centavos sea $\frac{1}{6}$? — 8 monedas de 25 centavos

19. En un juego se lanzan dos cubos numéricos. Para tener el, primer turno es necesario obtener un total de 6, 7 u 8. ¿Cuál es la probabilidad de que obtengas el primer turno? — $\frac{4}{9}$

Copyright © by Holt, Rinehart and Winston. All rights reserved.

LESSON **10-5**
Practice
Independent and Dependent Events

Determine if the events are dependent or independent.

1. choosing a tie and shirt from the closet — independent
2. choosing a month and tossing a coin — independent
3. rolling two fair number cubes once, then rolling them again if you received the same number on both number cubes on the first roll — dependent

An experiment consists of rolling a fair number cube and tossing a fair coin.

4. Find the probability of getting a 5 on the number cube and tails on the dime. — $\frac{1}{12}$
5. Find the probability of getting an even number on the number cube and heads on the dime. — $\frac{1}{4}$
6. Find the probability of getting a 2 or 3 on the number cube and heads on the dime. — $\frac{1}{6}$

A box contains 3 red marbles, 6 blue marbles, and 1 white marble. The marbles are selected at random, one at a time, and are not replaced. Find the probability.

7. $P(\text{blue and red})$ — $\frac{1}{5} = 0.2$
8. $P(\text{white and blue})$ — $\frac{1}{15} \approx 0.0\overline{6}$
9. $P(\text{red and white})$ — $\frac{1}{30} \approx 0.0\overline{3}$
10. $P(\text{red and white and blue})$ — $\frac{1}{40} = 0.025$
11. $P(\text{red and red and blue})$ — $\frac{1}{20} = 0.05$
12. $P(\text{red and blue and blue})$ — $\frac{1}{8} = 0.125$
13. $P(\text{red and red and red})$ — $\frac{1}{120} = 0.008\overline{3}$
14. $P(\text{white and blue and blue})$ — $\frac{1}{24} = 0.041\overline{6}$
15. $P(\text{white and red and white})$ — 0

Copyright © by Holt, Rinehart and Winston. All rights reserved.

LECCIÓN **10-5**
Práctica
Sucesos independientes y dependientes

Determina si los sucesos son dependientes o independientes.

1. elegir una corbata y una camisa del armario — independiente
2. elegir un mes y arrojar una moneda — independiente
3. lanzar dos dados una vez y luego volver a lanzarlos si la primera vez salió el mismo número en los dos dados — dependiente

Un experimento consiste en lanzar un dado y una moneda.

4. Halla la probabilidad de que salga un 5 en el dado y cruz en la moneda. — $\frac{1}{12}$
5. Halla la probabilidad de que salga un número par en el dado y cara en la moneda. — $\frac{1}{4}$
6. Halla la probabilidad de que salga un 2 o un 3 en el dado y cara en la moneda. — $\frac{1}{6}$

En una caja hay 3 canicas rojas, 6 canicas azules y 1 blanca. Las canicas se seleccionan al azar, de a una por vez, y no se reemplazan. Halla la probabilidad.

7. $P(\text{azul y rojo})$ — $\frac{1}{5} = 0.2$
8. $P(\text{blanco y azul})$ — $\frac{1}{15} \approx 0.0\overline{6}$
9. $P(\text{rojo y blanco})$ — $\frac{1}{30} \approx 0.0\overline{3}$
10. $P(\text{rojo, blanco y azul})$ — $\frac{1}{40} = 0.025$
11. $P(\text{rojo, rojo y azul})$ — $\frac{1}{20} = 0.05$
12. $P(\text{rojo, azul y azul})$ — $\frac{1}{8} = 0.125$
13. $P(\text{rojo, rojo y rojo})$ — $\frac{1}{120} = 0.008\overline{3}$
14. $P(\text{blanco, azul y azul})$ — $\frac{1}{24} = 0.041\overline{6}$
15. $P(\text{blanco, rojo y blanco})$ — 0

Copyright © by Holt, Rinehart and Winston. All rights reserved.

Copyright © by Holt, Rinehart and Winston. All rights reserved.

LESSON **10-6**

Practice

Making Decisions and Predictions

A sports store sells water bottles in different colors. The table shows the colors of the last 200 water bottles sold. The manager plans to order 1800 new water bottles.

Water Bottles Sold

Color	Number
Red	30
Blue	50
Green	25
Yellow	10
Purple	10
Clear	75

1. How many red water bottles should the manager order? 270

2. How many green water bottles should the manager order? 225

3. How many clear water bottles should the manager order? 675

4. If the carnival spinner lands on 10, the player gets a large stuffed animal. Suppose the spinner is spun 30 times. Predict how many large stuffed animals will be given away. 5

Decide whether the game is fair.

5. Roll two fair number cubes labeled 1–6. Player A wins if both numbers are the same. Player B wins if both numbers are different.

   not fair: $\frac{1}{6} \neq \frac{5}{6}$

6. Roll two fair number cubes labeled 1–6. Add the numbers. Player A wins if the sum is 5 or less. Player B wins if the sum is 9 or more.

   fair: $\frac{5}{18} = \frac{5}{18}$

7. Toss three fair coins. Player A wins if exactly one tail lands up. Otherwise, Player B wins.

   not fair: $\frac{3}{8} \neq \frac{5}{8}$

Copyright © by Holt, Rinehart and Winston. All rights reserved.

LECCIÓN **10-6**

Práctica

Cómo tomar decisiones y hacer predicciones

Una tienda de deportes vende botellas de agua de diferentes colores. En la tabla se muestran los colores de las últimas 200 botellas vendidas. El gerente piensa hacer un pedido de 1800 botellas de agua adicionales.

Botellas de agua vendidas

Color	Cantidad
Roja	30
Azul	50
Verde	25
Amarilla	10
Morada	10
Transparente	75

1. ¿Cuántas botellas de color rojo debe pedir el gerente? 270

2. ¿Cuántas botellas de color verde debe pedir el gerente? 225

3. ¿Cuántas botellas transparentes debe pedir el gerente? 675

4. Si la rueda de la fortuna cae en 10, el jugador se lleva un gran peluche. Supongamos que la rueda gira 30 veces. Predice cuántos peluches se entregarán. 5

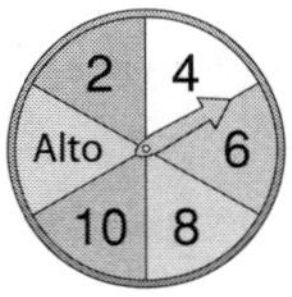

Decide si es un juego justo.

5. Se lanzan dos dados numerados del 1 al 6. El jugador A gana si saca dos números iguales. El jugador B gana si saca dos números distintos.

   no es justo: $\frac{1}{6} \neq \frac{5}{6}$

6. Se lanzan dos dados numerados del 1 al 6. Se suman los números. El jugador A gana si la suma da 5 o menos de 5. El jugador B gana si la suma da 9 o más.

   es justo: $\frac{5}{18} = \frac{5}{18}$

7. Se lanzan tres monedas. El jugador A gana sólo si cae una sola cruz. De no ser así, gana el jugador B.

   no es justo: $\frac{3}{8} \neq \frac{5}{8}$

Copyright © by Holt, Rinehart and Winston. All rights reserved.

LESSON **10-7**

Practice

Odds

A bag contains 9 red marbles, 5 green marbles, and 6 purple marbles.

1. Find *P*(red marble) $\frac{9}{20} = 0.45$

2. Find *P*(green marble) $\frac{1}{4} = 0.25$

3. Find *P*(purple marble) $\frac{3}{10} = 0.3$

4. Find the odds in favor of choosing a red marble.

   9:11

5. Find the odds against choosing a red marble.

   11:9

6. Find the odds in favor of choosing a green marble.

   5:15 = 1:3

7. Find the odds against choosing a green marble.

   15:5 = 3:1

8. Find the odds in favor of choosing a purple marble.

   6:14 = 3:7

9. Find the odds against choosing a purple marble.

   14:6 = 7:3

10. Find the odds in favor of not choosing a green marble.

    15:5 = 3:1

11. Find the odds in favor of choosing a red or purple marble.

    15:5 = 3:1

12. If the probability of Helena winning the contest is $\frac{2}{5}$, what are the odds in favor of Helena winning the contest?

    2:3

13. The odds in favor of the Bruins winning the Stanley Cup are 5 to 4. What is the probability that the Bruins will win the Stanley Cup?

    $\frac{5}{9} \approx 0.55\overline{5}$

Copyright © by Holt, Rinehart and Winston. All rights reserved.

LECCIÓN **10-7**

Práctica

Probabilidades a favor y en contra

En una bolsa hay 9 canicas rojas, 5 canicas verdes y 6 moradas.

1. Halla *P*(canica roja) $\frac{9}{20} = 0.45$

2. Halla *P*(canica verde) $\frac{1}{4} = 0.25$

3. Halla *P*(canica morada) $\frac{3}{10} = 0.3$

4. Halla las probabilidades a favor de que salga una canica roja.

   9:11

5. Halla las probabilidades en contra de que salga una canica roja.

   11:9

6. Halla las probabilidades a favor de que salga una canica verde.

   5:15 = 1:3

7. Halla las probabilidades en contra de que salga una canica verde.

   15:5 = 3:1

8. Halla las probabilidades a favor de que salga una canica morada.

   6:14 = 3:7

9. Halla las probabilidades en contra de que salga una canica morada.

   14:6 = 7:3

10. Halla las probabilidades a favor de que no salga una canica verde.

    15:5 = 3:1

11. Halla las probabilidades a favor de que salga una canica roja o morada.

    15:5 = 3:1

12. Si la probabilidad de que Helena gane el concurso es de $\frac{2}{5}$, ¿cuáles son las probabilidades a favor de que Helena gane el concurso?

    2:3

13. Las probabilidades de que Bruins gane la Copa Stanley son de 5 a 4. ¿Cuál es la probabilidad de que Bruins gane la Copa Stanley?

    $\frac{5}{9} \approx 0.55\overline{5}$

Copyright © by Holt, Rinehart and Winston. All rights reserved.

Copyright © by Holt, Rinehart and Winston.
All rights reserved.

LESSON 10-8

Practice
Counting Principles

Employee identification codes at a company contain 2 letters followed by 2 numbers. All codes are equally likely.

1. Find the number of possible identification codes.

   67,600

2. Find the probability of being assigned the code MT49.

   $\frac{1}{67,600} \approx 0.000015$

3. Find the probability that an ID code of the company does not contain the letter *A* as the second letter of the code.

   $\frac{65,000}{67,600} = \frac{25}{26} \approx 0.962$

4. Find the probability that an ID code of the company does not contain the number 2.

   $\frac{54,756}{67,600} = \frac{81}{100} = 0.81$

5. Mrs. Sharpe is planning her dinners for next week. The choices for the entree are roast beef, turkey, or pork. The choices of carbohydrates are mashed potatoes, baked potatoes, or noodles. The vegetable choices are broccoli, spinach, or carrots. Make a tree diagram indicating the possible outcomes for each entree.

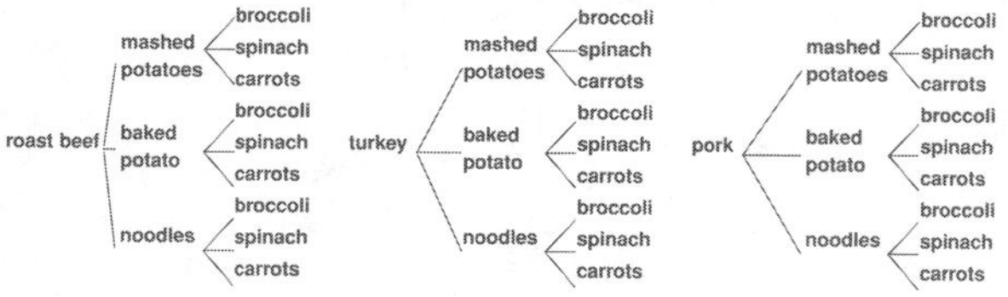

6. How many different meals could Mrs. Sharpe prepare? 27

Find the probability for each of the following.

7. *P*(dinner with baked potato)

   $\frac{1}{3} = 0.33\overline{3}$

8. *P*(dinner with noodles and carrots)

   $\frac{1}{9} = 0.11\overline{1}$

9. Mitch bought 2 sports magazines, 3 guitar magazines, and 3 news magazines. How many choices of magazines does he have to read?

   8

Copyright © by Holt, Rinehart and Winston. All rights reserved.

LECCIÓN 10-8

Práctica
Principios de conteo

Los códigos de identificación de los empleados de una compañía consisten en 2 letras seguidas de 2 números. Todos los códigos son igualmente probables.

1. Halla la cantidad de códigos de identificación posibles.

   67,600

2. Halla la probabilidad de que te asignen el código MT49.

   $\frac{1}{67,600} \approx 0.000015$

3. Halla la probabilidad de que un código de identificación de la compañía no tenga a la letra *A* en la segunda posición del código.

   $\frac{65,000}{67,600} = \frac{25}{26} \approx 0.962$

4. Halla la probabilidad de que un código de identificación de la compañía no tenga el número 2.

   $\frac{54,756}{67,600} = \frac{81}{100} = 0.81$

5. La señora Sharpe está planificando las cenas de la próxima semana. Las opciones para la entrada son rosbif, pavo o cerdo. Las opciones de carbohidratos son puré de papas, papas al horno o fideos. Las opciones de verduras son brócolis, espinacas o zanahorias. Haz un diagrama de árbol que indique los posibles resultados de cada entrada.

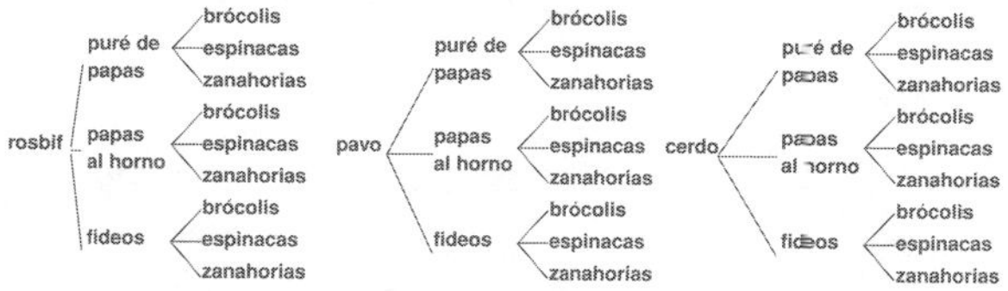

6. ¿Cuántas comidas distintas puede preparar la señora Sharpe? 27

Halla la probabilidad de cada uno de los siguientes casos.

7. *P*(cena con papas al horno)

   $\frac{1}{3} = 0.33\overline{3}$

8. *P*(cena con fideos y zanahorias)

   $\frac{1}{9} = 0.11\overline{1}$

9. Mitch compró 2 revistas de deportes, 3 revistas de guitarra y 3 revistas de noticias. ¿Cuántas opciones de revistas puede leer?

   8

Copyright © by Holt, Rinehart and Winston. All rights reserved.

LESSON 10-9

Practice
Permutations and Combinations

Evaluate each expression.

1. 10!

   3,628,800

2. 13!

   6,227,020,800

3. 11! − 8!

   39,876,480

4. 12! − 9!

   478,638,720

5. $\frac{15!}{8!}$

   32,432,400

6. $\frac{18!}{12!}$

   13,366,080

7. $\frac{13!}{(17-12)!}$

   51,891,840

8. $\frac{19!}{(15-2)!}$

   19,535,040

9. $\frac{15!}{(18-10)!}$

   32,432,400

10. Signaling is a means of communication through signals or objects. During the time of the American Revolution, the colonists used combinations of a barrel, basket, and a flag placed in different positions atop a post. How many different signals could be sent by using 3 flags, one above the other on a pole, if 8 different flags were available?

    336

11. From a class of 25 students, how many different ways can 4 students be selected to serve in a mock trial as the judge, defending attorney, prosecuting attorney, and the defendant?

    303,600

12. How many different 4 people committees can be formed from a group of 15 people?

    1365

13. The girls' basketball team has 12 players. If the coach chooses 5 girls to play at a time, how many different teams can be formed?

    792

14. A photographer has 50 pictures to be placed in an album. How many combinations will the photographer have to choose from if there will be 6 pictures placed on the first page?

    15,890,700

Copyright © by Holt, Rinehart and Winston. All rights reserved.

LECCIÓN 10-9

Práctica
Permutaciones y combinaciones

Evalúa cada expresión.

1. 10!

   3,628,800

2. 13!

   6,227,020,800

3. 11! − 8!

   39,876,480

4. 12! − 9!

   478,638,720

5. $\frac{15!}{8!}$

   32,432,400

6. $\frac{18!}{12!}$

   13,366,080

7. $\frac{13!}{(17-12)!}$

   51,891,840

8. $\frac{19!}{(15-2)!}$

   19,535,040

9. $\frac{15!}{(18-10)!}$

   32,432,400

10. La señalización es un método para comunicarse a través de señales u objetos. En los tiempos de la Revolución estadounidense, los colonos emplearon combinaciones de un barril, una cesta y una bandera ubicados en distintas posiciones encima de un poste. ¿Cuántas señales distintas podían enviarse con 3 banderas, una sobre la otra en un poste, si había 8 banderas distintas disponibles?

    336

11. De una clase de 25 estudiantes, ¿de cuántas formas distintas se puede elegir a 4 estudiantes para que hagan las veces de juez, abogado defensor, fiscal y acusado?

    303,600

12. ¿Cuántas comisiones con 4 integrantes distintos se pueden formar con un grupo de 15 personas?

    1365

13. El equipo femenino de básquetbol tiene 12 jugadoras. Si el entrenador elige 5 chicas para que jueguen por vez, ¿cuántos equipos distintos se pueden armar?

    792

14. Un fotógrafo tiene 50 fotografías para colocar en un álbum. ¿Entre cuántas combinaciones deberá elegir el fotógrafo si se colocan 6 fotografías en la primera página?

    15,890,700

Copyright © by Holt, Rinehart and Winston. All rights reserved.

Copyright © by Holt, Rinehart and Winston. All rights reserved.

LESSON **11-1**
Practice
Simplifying Algebraic Expressions

Combine like terms.

1. $8a - 5a$ — $3a$
2. $12g + 7g$ — $19g$
3. $4a + 7a + 6$ — $11a + 6$
4. $6x + 3y + 5x$ — $11x + 3y$
5. $10k - 3k + 5h$ — $7k + 5h$
6. $3p - 7q + 14p$ — $17p - 7q$
7. $3k + 7k + 5k$ — $15k$
8. $5c + 12d - 6$ — $5c + 12d - 6$
9. $13 + 4b + 6b - 5$ — $8 + 10b$
10. $4f + 6 + 7f - 2$ — $11f + 4$
11. $x + y + 3x + 7y$ — $4x + 8y$
12. $9n + 13 - 8n - 6$ — $n + 7$

Simplify.

13. $4(x + 3) - 5$ — $4x + 7$
14. $6(7 + x) + 5x$ — $42 + 11x$
15. $3(5 + 3x) - 4x$ — $15 + 5x$

Solve.

16. $6y + 2y = 16$ — $y = 2$
17. $14b - 9b = 35$ — $b = 7$
18. $3q + 9q = 48$ — $q = 4$

19. Gregg has q quarters and p pennies. His brother has 4 times as many quarters and 8 times as many pennies as Gregg has. Write the sum of the number of coins they have, and then combine like terms.

$q + p + 4q + 8p$; $5q + 9p$

20. If Gregg has 6 quarters and 15 pennies, how many total coins do Gregg and his brother have?

165 coins

Copyright © by Holt, Rinehart and Winston. All rights reserved.

LECCIÓN **11-1**
Práctica
Cómo simplificar expresiones algebraicas

Combina los términos semejantes.

1. $8a - 5a$ — $3a$
2. $12g + 7g$ — $19g$
3. $4a + 7a + 6$ — $11a + 6$
4. $6x + 3y + 5x$ — $11x + 3y$
5. $10k - 3k + 5h$ — $7k + 5h$
6. $3p - 7q + 14p$ — $17p - 7q$
7. $3k + 7k + 5k$ — $15k$
8. $5c + 12d - 6$ — $5c + 12d - 6$
9. $13 + 4b + 6b - 5$ — $8 + 10b$
10. $4f + 6 + 7f - 2$ — $11f + 4$
11. $x + y + 3x + 7y$ — $4x + 8y$
12. $9n + 13 - 8n - 6$ — $n + 7$

Simplifica.

13. $4(x + 3) - 5$ — $4x + 7$
14. $6(7 + x) + 5x$ — $42 + 11x$
15. $3(5 + 3x) - 4x$ — $15 + 5x$

Resuelve.

16. $6y + 2y = 16$ — $y = 2$
17. $14b - 9b = 35$ — $b = 7$
18. $3q + 9q = 48$ — $q = 4$

19. Gregg tiene q monedas de 25 centavos y p monedas de 1 centavo. Su hermano tiene el cuádruple de monedas de 25 centavos y 8 veces la cantidad de monedas de 1 centavo que tiene Gregg. Escribe la suma de la cantidad de monedas que tienen y combina los términos semejantes.

$q + p + 4q + 8p$; $5q + 9p$

20. Si Gregg tiene 6 monedas de 25 centavos y 15 monedas de 1 centavo, ¿cuántas monedas tienen en total Gregg y su hermano?

165 monedas

Copyright © by Holt, Rinehart and Winston. All rights reserved.

LESSON **11-2**
Practice
Solving Multi-Step Equations

Solve.

1. $2x + 5x + 4 = 25$ — $x = 3$
2. $9 + 3y - 2y = 14$ — $y = 5$
3. $16 = 4w + 2w - 2$ — $w = 3$
4. $26 = 3b - 2 - 7b$ — $b = -7$
5. $31 + 4t - t = 40$ — $t = 3$
6. $14 - 2x + 4x = 20$ — $x = 3$
7. $\frac{5m}{8} - \frac{6}{8} + \frac{3m}{8} = \frac{2}{8}$ — $m = 1$
8. $-4\frac{2}{3} = \frac{2n}{3} + \frac{1}{3} + \frac{n}{3}$ — $n = -5$
9. $7a + 16 - 3a = -4$ — $a = -5$
10. $\frac{x}{2} + 1 + \frac{3x}{4} = -9$ — $x = -8$
11. $7m + 3 - 4m = -9$ — $m = -4$
12. $\frac{2x}{5} + 3 - \frac{4x}{5} = \frac{1}{5}$ — $x = 7$
13. $\frac{7k}{8} - \frac{3}{4} - \frac{5k}{16} = \frac{3}{8}$ — $k = 2$
14. $6y + 9 - 4y = -3$ — $y = -6$
15. $\frac{5a}{6} - \frac{7}{12} + \frac{3a}{4} = -2\frac{1}{6}$ — $a = -1$

16. The measure of an angle is 28° greater than its complement. Find the measure of each angle.

angle = 59°; complement = 31°

17. The measure of an angle is 21° more than twice its supplement. Find the measure of each angle.

angle = 127°; supplement = 53°

18. The perimeter of the triangle is 126 units. Find the measure of each side.

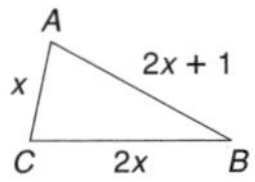

AC = 25 units; BC = 50 units; AB = 51 units

19. The base angles of an isosceles triangle are congruent. If the measure of each of the base angles is twice the measure of the third angle, find the measure of all three angles.

36°; 72°; 72°

Copyright © by Holt, Rinehart and Winston. All rights reserved.

LECCIÓN **11-2**
Práctica
Cómo resolver ecuaciones de varios pasos

Resuelve.

1. $2x + 5x + 4 = 25$ — $x = 3$
2. $9 + 3y - 2y = 14$ — $y = 5$
3. $16 = 4w + 2w - 2$ — $w = 3$
4. $26 = 3b - 2 - 7b$ — $b = -7$
5. $31 + 4t - t = 40$ — $t = 3$
6. $14 - 2x + 4x = 20$ — $x = 3$
7. $\frac{5m}{8} - \frac{6}{8} + \frac{3m}{8} = \frac{2}{8}$ — $m = 1$
8. $-4\frac{2}{3} = \frac{2n}{3} + \frac{1}{3} + \frac{n}{3}$ — $n = -5$
9. $7a + 16 - 3a = -4$ — $a = -5$
10. $\frac{x}{2} + 1 + \frac{3x}{4} = -9$ — $x = -8$
11. $7m + 3 - 4m = -9$ — $m = -4$
12. $\frac{2x}{5} + 3 - \frac{4x}{5} = \frac{1}{5}$ — $x = 7$
13. $\frac{7k}{8} - \frac{3}{4} - \frac{5k}{16} = \frac{3}{8}$ — $k = 2$
14. $6y + 9 - 4y = -3$ — $y = -6$
15. $\frac{5a}{6} - \frac{7}{12} + \frac{3a}{4} = -2\frac{1}{6}$ — $a = -1$

16. Un ángulo mide 28° más que su complemento. Halla la medida de cada ángulo.

ángulo = 59°; complemento = 31°

17. Un ángulo tiene 21° más que el doble de su suplemento. Halla la medida de cada ángulo.

ángulo = 127°; suplemento = 53°

18. El perímetro del triángulo mide 126 unidades. Halla la medida de cada lado.

AC = 25 unidades; BC = 50 unidades; AB = 51 unidades

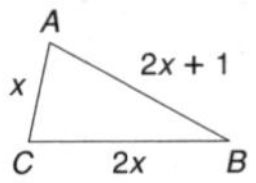

19. Los ángulos de la base de un triángulo isósceles son congruentes. Si cada uno de los ángulos de la base mide el doble que el tercer ángulo, halla las medidas de los tres ángulos.

36°; 72°; 72°

Copyright © by Holt, Rinehart and Winston. All rights reserved.

Copyright © by Holt, Rinehart and Winston.
All rights reserved.

LESSON **11-3**

Practice

Solving Equations with Variables on Both Sides

Solve.

1. $7x - 11 = -19 + 3x$ — $x = -2$
2. $11a + 9 = 4a + 30$ — $a = 3$
3. $4t + 14 = \frac{6t}{5} + 7$ — $t = -2.5$
4. $19c + 31 = 26c - 74$ — $c = 15$
5. $\frac{3y}{8} - 9 = 13 + \frac{y}{8}$ — $y = 88$
6. $\frac{3k}{5} + 44 = \frac{12k}{25} + 8$ — $k = -300$
7. $10a - 37 = 6a + 51$ — $a = 22$
8. $5w + 9.9 = 4.8 + 8w$ — $w = 1.7$
9. $15 - x = 2(x + 3)$ — $x = 3$
10. $15y + 14 = 2(5y + 6)$ — $y = -0.4$
11. $14 - \frac{w}{8} = \frac{3w}{4} - 21$ — $w = 40$
12. $\frac{1}{2}(6x - 4) = 4x - 9$ — $x = 7$
13. $4(3d - 2) = 8d - 5$ — $d = \frac{3}{4}$
14. $\frac{y}{3} + 11 = \frac{y}{2} - 3$ — $y = 84$
15. $\frac{2x - 9}{3} = 8 - 3x$ — $x = 3$

16. Forty-eight decreased by a number is the same as the difference of four times the number and seven. Find the number.

 11

17. The square and the equilateral triangle at the right have the same perimeter. Find the length of the sides of the triangle.

 12 units

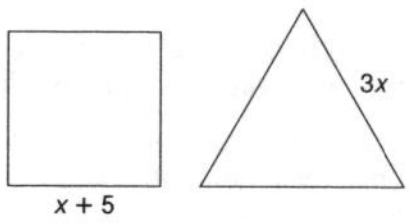

LECCIÓN **11-3**

Práctica

Cómo resolver ecuaciones con variables en ambos lados

Resuelve.

1. $7x - 11 = -19 + 3x$ — $x = -2$
2. $11a + 9 = 4a + 30$ — $a = 3$
3. $4t + 14 = \frac{6t}{5} + 7$ — $t = -2.5$
4. $19c + 31 = 26c - 74$ — $c = 15$
5. $\frac{3y}{8} - 9 = 13 + \frac{y}{8}$ — $y = 88$
6. $\frac{3k}{5} + 44 = \frac{12k}{25} + 8$ — $k = -300$
7. $10a - 37 = 6a + 51$ — $a = 22$
8. $5w + 9.9 = 4.8 + 8w$ — $w = 1.7$
9. $15 - x = 2(x + 3)$ — $x = 3$
10. $15y + 14 = 2(5y + 6)$ — $y = -0.4$
11. $14 - \frac{w}{8} = \frac{3w}{4} - 21$ — $w = 40$
12. $\frac{1}{2}(6x - 4) = 4x - 9$ — $x = 7$
13. $4(3d - 2) = 8d - 5$ — $d = \frac{3}{4}$
14. $\frac{y}{3} + 11 = \frac{y}{2} - 3$ — $y = 84$
15. $\frac{2x - 9}{3} = 8 - 3x$ — $x = 3$

16. El resultado de la diferencia entre 48 y cierto número es igual al resultado de la diferencia entre el cuádruple de dicho número y 7. Halla el número.

 11

17. El cuadrado y el triángulo equilátero que se encuentran a la derecha tienen el mismo perímetro. Halla la longitud de los lados del triángulo.

 12 unidades

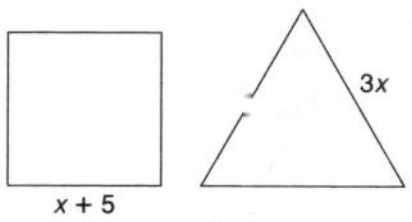

LESSON **11-4**

Practice

Solving Inequalities by Multiplying or Dividing

Solve and graph.

1. $\frac{m}{-5} \le 4$ — $m \ge -20$

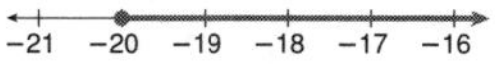

2. $-16 < -8n$ — $n < 2$

3. $7p \ge 49$ — $p \ge 7$

4. $10 > \frac{q}{2}$ — $q < 20$

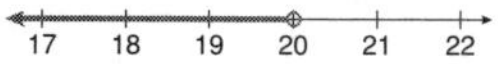

5. $-\frac{r}{3} \le 15$ — $r \ge -45$

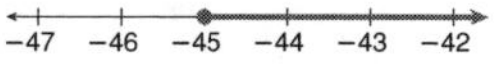

6. $22 > -2s$ — $s > -11$

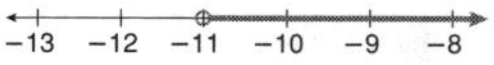

7. $-6t < -24$ — $t > 4$

8. $\frac{v}{20} \ge 2$ — $v \ge 40$

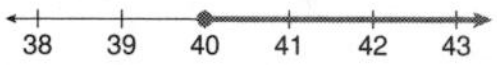

9. On a snorkeling trip, Antonia dove at least 7 times as deep as Lucy did. If Antonia dove 35 feet below the ocean's surface, what was the deepest that Lucy dove?

 5 feet

10. Last week, Saul ran more than one-fifth the distance that his friend Omar ran. If Saul ran 14 miles last week, how far did Omar run?

 less than 70 miles

LECCIÓN **11-4**

Práctica

Cómo resolver desigualdades mediante la multiplicación o la división

Resuelve y representa gráficamente.

1. $\frac{m}{-5} \le 4$ — $m \ge -20$

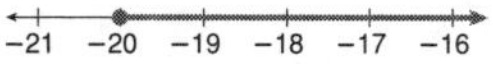

2. $-16 < -8n$ — $n < 2$

3. $7p \ge 49$ — $p \ge 7$

4. $10 > \frac{q}{2}$ — $q < 20$

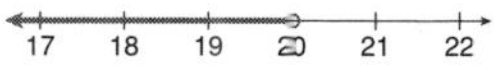

5. $-\frac{r}{3} \le 15$ — $r \ge -45$

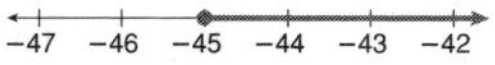

6. $22 > -2s$ — $s > -11$

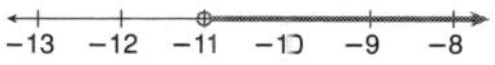

7. $-6t < -24$ — $t > 4$

8. $\frac{v}{20} \ge 2$ — $v \ge 40$

9. En una excursión de buceo, Antonia se sumergió a una profundidad al menos 7 veces mayor que la de Lucy. Si Antonia se sumergió 35 pies bajo la superficie del océano, ¿cuál fue la profundidad máxima a la que descendió Lucy?

 5 pies

10. La semana pasada, Saúl corrió más de un quinto de la distancia que recorrió su amigo Omar. Si Saúl corrió 14 millas la semana pasada, ¿cuánto corrió Omar?

 menos de 70 millas

LESSON **11-5**

Practice

Solving Two-Step Inequalities

Solve and graph.

1. $4x - 2 < 26$

   $x < 7$

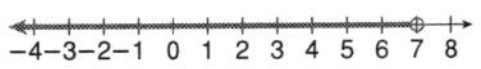

2. $6 - \frac{1}{5}y \le 7$

   $y \ge -5$

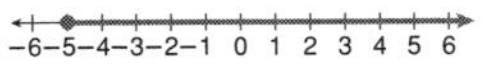

3. $2x + 27 \ge 15$

   $x \ge -6$

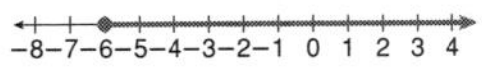

4. $10x > 14x + 8$

   $x < -2$

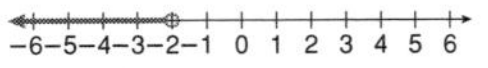

5. $7 - 4w \le 19$

   $w \ge -3$

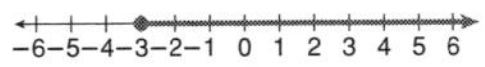

6. $\frac{k}{5} + \frac{3}{20} < \frac{3}{10}$

   $k < \frac{3}{4}$

7. $4.8 - 9.6x \le 14.4$

   $x \ge -1$

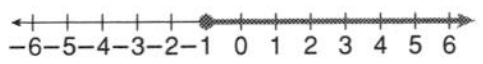

8. $\frac{2}{9} + \frac{y}{3} > \frac{1}{3}$

   $y > \frac{1}{3}$

9. One-third of a number, decreased by thirty-six, is at most twenty-two. Find the number. $n \le 174$

10. Jack wants to run at least 275 miles before the baseball season begins. He has already run 25 miles. He plans to run 2.5 miles each day. At this rate, what is the fewest number of days he will need to reach his goal? 100 days

Copyright © by Holt, Rinehart and Winston. All rights reserved.

LECCIÓN **11-5**

Práctica

Cómo resolver desigualdades de dos pasos

Resuelve y representa gráficamente.

1. $4x - 2 < 26$

   $x < 7$

2. $6 - \frac{1}{5}y \le 7$

   $y \ge -5$

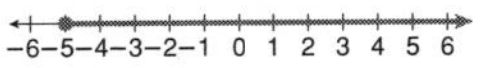

3. $2x + 27 \ge 15$

   $x \ge -6$

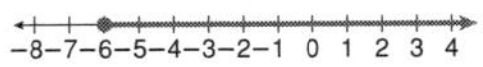

4. $10x > 14x + 8$

   $x < -2$

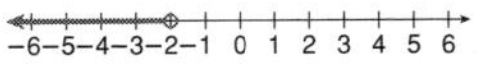

5. $7 - 4w \le 19$

   $w \ge -3$

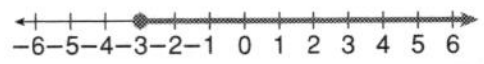

6. $\frac{k}{5} + \frac{3}{20} < \frac{3}{10}$

   $k < \frac{3}{4}$

7. $4.8 - 9.6x \le 14.4$

   $x \ge -1$

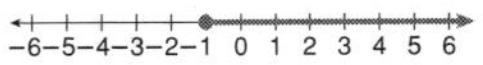

8. $\frac{2}{9} + \frac{y}{3} > \frac{1}{3}$

   $y > \frac{1}{3}$

9. La diferencia entre el tercio de un número y 36 es al menos 22. Halla el número. $n \le 174$

10. Jack quiere correr al menos 275 millas antes de que comience la temporada de béisbol. Ya ha corrido 25 millas. Piensa correr 2.5 millas por día. A este ritmo, ¿cuál es la menor cantidad de días que va a necesitar para alcanzar su objetivo? 100 semanas

Copyright © by Holt, Rinehart and Winston. All rights reserved.

LESSON **11-6**

Practice

Systems of Equations

Solve each system of equations.

1. $y = 2x - 4$
   $y = x - 1$
   (3,2)

2. $y = -x + 10$
   $y = x + 2$
   (4,6)

3. $y = 2x - 1$
   $y = -3x - 6$
   (−1, −3)

4. $y = 2x$
   $y = 12 - x$
   (4,8)

5. $y = 2x - 3$
   $y = 2x + 1$
   no solution

6. $y = 3x - 1$
   $y = x + 1$
   (1,2)

7. $x + y = 0$
   $5x + 2y = -3$
   (−1,1)

8. $2x - 3y = 0$
   $2x + y = 8$
   (3,2)

9. $2x + 3y = 6$
   $4x + 6y = 12$
   infinite number

10. $6x - y = -14$
    $2x - 3y = 6$
    (−3, −4)

11. The sum of two numbers is 24. The second number is 6 less than the first. Write a system of equations and solve it find the number.

    $x + y = 24$; $y = x - 6$; (15, 9)

15. Kerry and Luke biked a total of 18 miles in one weekend. Kerry biked 4 miles more than Luke. Write a system of equations and solve it to find how far each boy biked.

    $x + y = 18$; $x + 4 = y$; (7, 11); Luke biked 7 miles, and Kerry biked 11 miles

Copyright © by Holt, Rinehart and Winston. All rights reserved.

LECCIÓN **11-6**

Práctica

Sistemas de ecuaciones

Resuelve cada sistema de ecuaciones.

1. $y = 2x - 4$
   $y = x - 1$
   (3,2)

2. $y = -x + 10$
   $y = x + 2$
   (4,6)

3. $y = 2x - 1$
   $y = -3x - 6$
   (−1, −3)

4. $y = 2x$
   $y = 12 - x$
   (4,8)

5. $y = 2x - 3$
   $y = 2x + 1$
   sin solución

6. $y = 3x - 1$
   $y = x + 1$
   (1,2)

7. $x + y = 0$
   $5x + 2y = -3$
   (−1,1)

8. $2x - 3y = 0$
   $2x + y = 8$
   (3,2)

9. $2x + 3y = 6$
   $4x + 6y = 12$
   número infinito

10. $6x - y = -14$
    $2x - 3y = 6$
    (−3, −4)

11. La suma de dos números da 24. El segundo número es 6 unidades menor que el primero. Escribe un sistema de ecuaciones y resuélvelo para hallar el número.

    $x + y = 24$; $y = x - 6$; (15, 9)

15. Kerry y Luke recorrieron en bicicleta un total de 18 millas en un fin de semana. Kerry recorrió 4 millas más que Luke. Escribe un sistema de ecuaciones y resuélvelo para hallar la distancia que recorrió cada uno.

    $x + y = 18$; $x + 4 = y$; (7, 11); Luke recorrió 7 millas y Kerry, 11 millas

Copyright © by Holt, Rinehart and Winston. All rights reserved.

Copyright © by Holt, Rinehart and Winston. All rights reserved.

LESSON **12-1**
Practice
Graphing Linear Equations

Graph each equation and tell whether it is linear.

1. $y = -2x - 5$

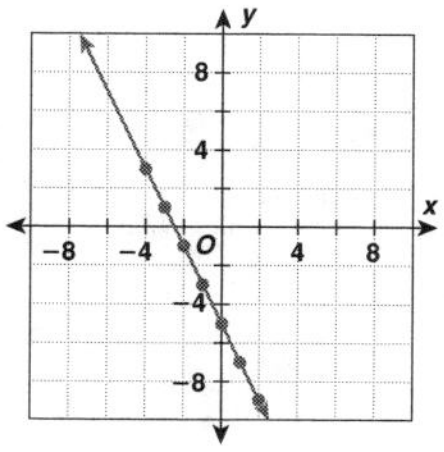

linear

2. $y = -x^2 + 1$

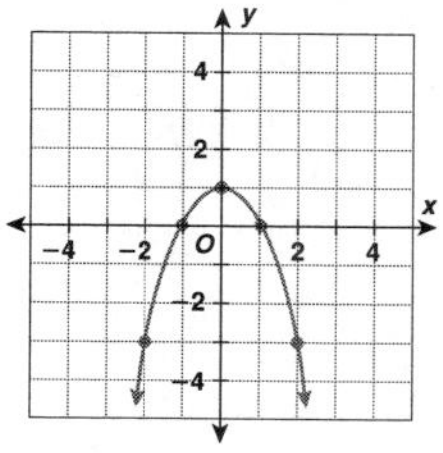

not linear

3. $y = x^2 - 7$

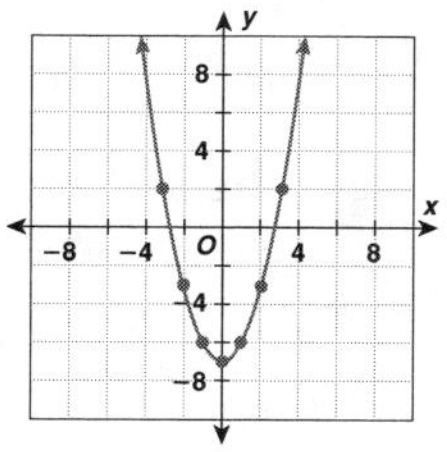

not linear

4. $y = \frac{1}{2}x - 1$

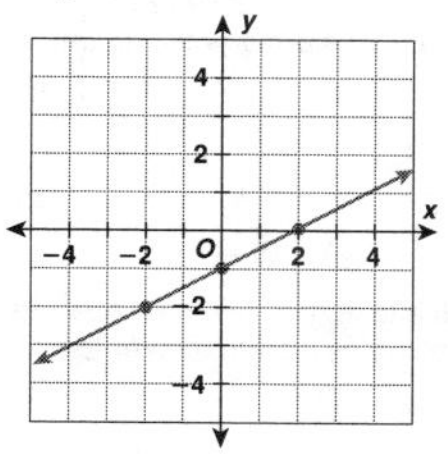

linear

5. A real estate agent commission may be based on the equation $C = 0.06s + 450$, where s represents the total sales. If the agent sells a property for $125,000, what is the commission earned by the agent? Graph the equation and tell whether it is linear.

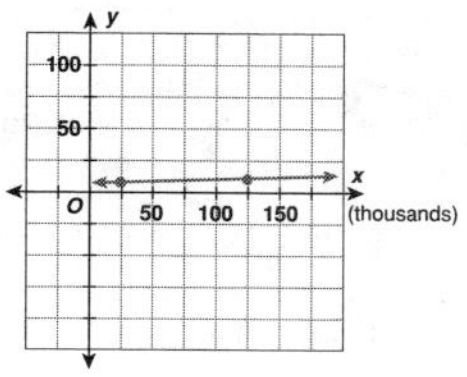

$7950; linear

Copyright © by Holt, Rinehart and Winston.
All rights reserved.

LECCIÓN **12-1**
Práctica
Cómo representar gráficamente ecuaciones lineales

Representa gráficamente cada ecuación e indica si es lineal.

1. $y = -2x - 5$

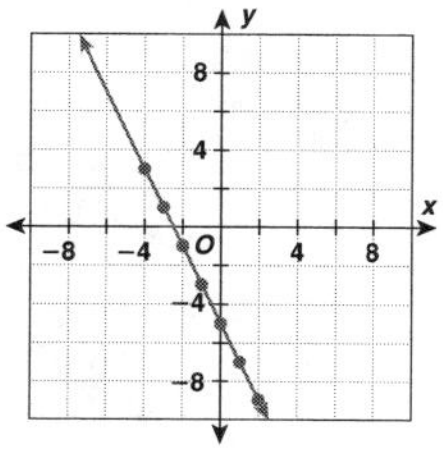

lineal

2. $y = -x^2 + 1$

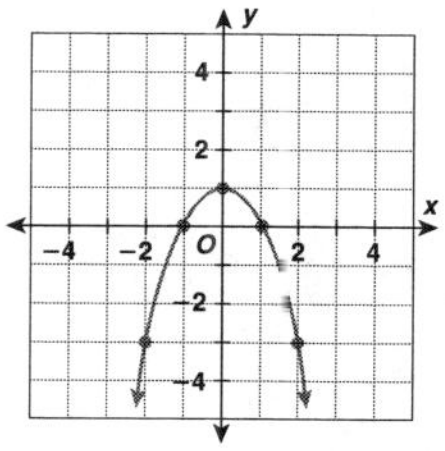

no lineal

3. $y = x^2 - 7$

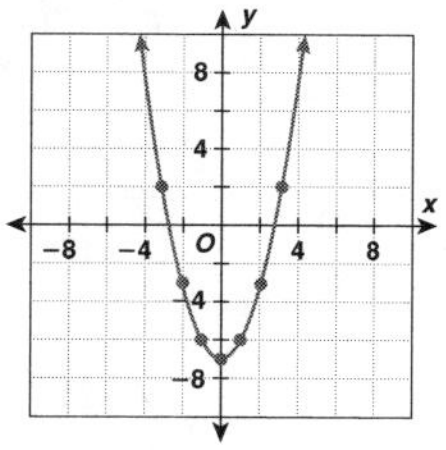

no lineal

4. $y = \frac{1}{2}x - 1$

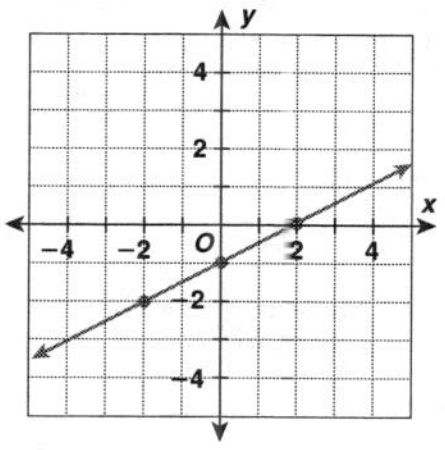

lineal

5. La comisión de un agente de bienes raíces puede basarse en la ecuación $C = 0.06v + 450$, donde v representa las ventas totales. Si el agente vende una propiedad en $125,000, ¿qué comisión le corresponde? Representa gráficamente la ecuación e indica si es lineal.

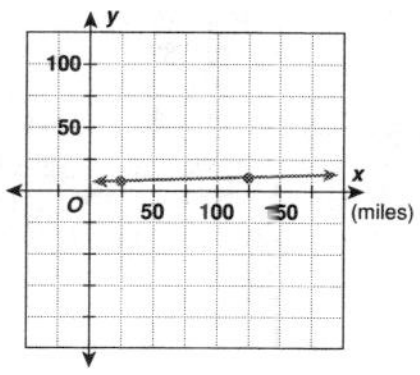

$7950; lineal

Copyright © by Holt, Rinehart and Winston.
All rights reserved.

LESSON **12-2**
Practice
Slope of a Line

Find the slope of the line that passes through each pair of points.

1. (−2, −8), (1, 4) — 4
2. (−2, 0), (0, 4), — 2
3. (0, 4), (4, 4) — 0
4. (3, −6), (2, −4) — −2
5. (−3, 4), (3, −4) — $-\frac{4}{3}$
6. (3, 0), (0, −6), — 2
7. (3, 2), (3, −2) — undefined
8. (−4, 4), (3, −1) — $-\frac{5}{7}$

Determine whether each graph shows a constant or variable rate of change. Explain your reasoning.

9.

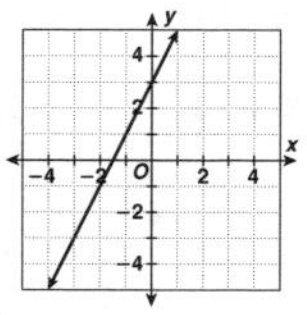

constant; The slope between any two points is always the same.

10.

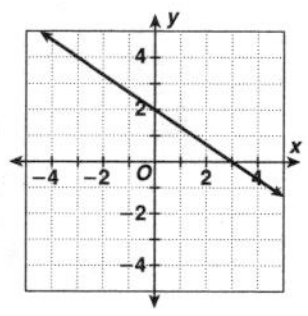

constant; The slope between any two points is always the same.

11.

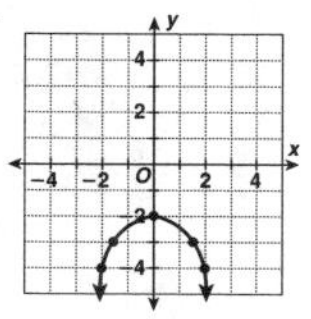

variable; The slope is positive in Quadrant III and negative in Quadrant IV.

12. The table shows the distance Ms. Long had traveled as she went to the beach. Use the data to make a graph. Find the slope of the line and explain what it shows.

Time (min)	Distance (mi)
8	6
12	9
16	12
20	15

The slope is $\frac{3}{4}$, which means that for every 4 minutes Ms. Long drives, she travels 3 miles. She is driving 45 mph.

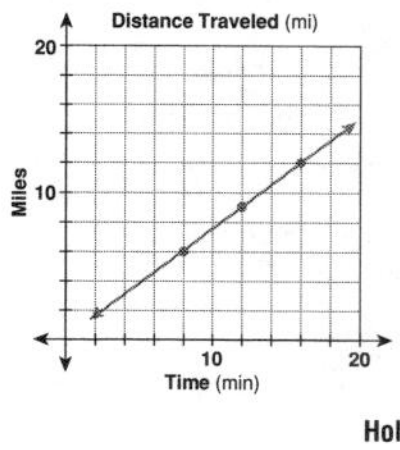

Copyright © by Holt, Rinehart and Winston.
All rights reserved.

LECCIÓN **12-2**
Práctica
Pendiente de una línea

Halla la pendiente de la línea que pasa por cada par de puntos.

1. (−2, −8), (1, 4) — 4
2. (−2, 0), (0, 4), — 2
3. (0, 4), (4, 4) — 0
4. (3, −6), (2, −4) — −2
5. (−3, 4), (3, −4) — $-\frac{4}{3}$
6. (3, 0), (0, −6), — 2
7. (3, 2), (3, −2) — indefinida
8. (−4, 4), (3, −1) — $-\frac{5}{7}$

Determina si en cada gráfica se muestra una tasa de cambio constante o variable. Explica tu razonamiento.

9.

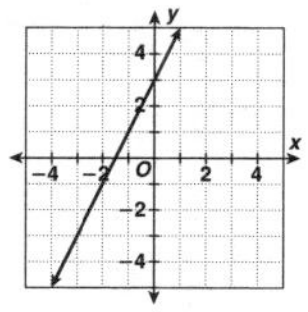

Constante. La pendiente entre dos puntos cualesquiera es siempre la misma.

10.

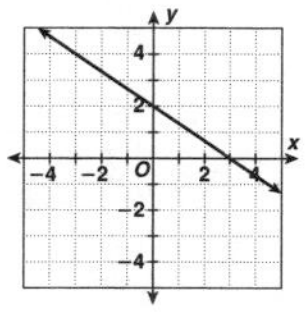

Constante. La pendiente entre dos puntos cualesquiera es siempre la misma.

11.

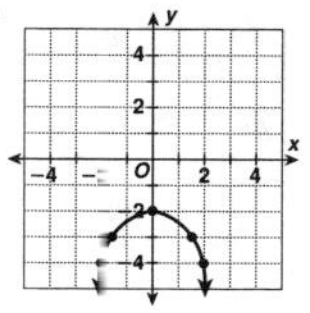

Variable. La pendiente es positiva en el Cuadrante III y negativa en el Cuadrante IV.

12. En la tabla se muestra la distancia que recorrió la señorita Long cuando fue a la playa. Usa los datos para hacer una gráfica. Halla la pendiente de la línea y explica lo que muestra.

Tiempo (min)	Distancia (mi)
8	6
12	9
16	12
20	15

La pendiente es de $\frac{3}{4}$, lo que significa que cada 4 minutos la señorita Long recorre 3 millas. Su velocidad es de 45 mph.

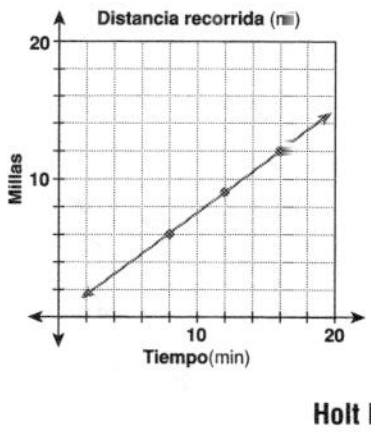

Copyright © by Holt, Rinehart and Winston.
All rights reserved.

Copyright © by Holt, Rinehart and Winston.
All rights reserved.

LESSON **12-3**

Practice

Using Slopes and Intercepts

Find the x-intercept and y-intercept of each line. Use the intercepts to graph the equation.

1. $x - y = -3$

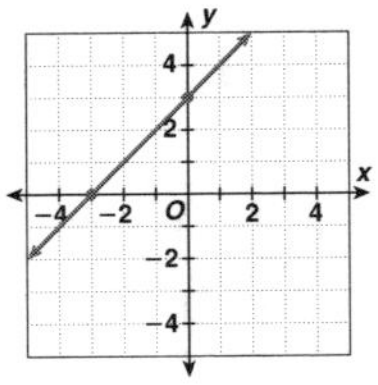

x-intercept is -3; y-intercept is 3

2. $2x + 3y = 12$

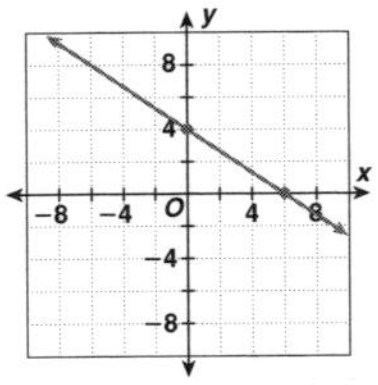

x-intercept is 6; y-intercept is 4

Write each equation in slope-intercept form, and then find the slope and y-intercept.

3. $3x + y = 0$

$y = -3x$; $m = -3$; $b = 0$

4. $2x - y = -15$

$y = 2x + 15$; $m = 2$; $b = 15$

5. $x - 5y = 10$

$y = \frac{1}{5}x - 2$; $m = \frac{1}{5}$; $b = -2$

Write the equation of the line that passes through each pair of points in slope-intercept form.

6. (3, 4), (4, 6)

$y = 2x - 2$

7. (−1, −1), (2, −10)

$y = -3x - 4$

8. (6, 5), (−9, −20)

$y = \frac{5}{3}x - 5$

9. A pizzeria charges $8 for a large cheese pizza, plus $2 for each topping. The total cost for a large pizza is given by the equation $C = 2t + 8$, where t is the number of toppings. Identify the slope and y-intercept, and use them to graph the equation for t between 0 and 5 toppings.

The slope is 2, and the y-intercept is 8.

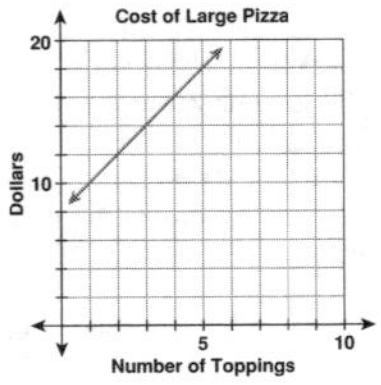

Copyright © by Holt, Rinehart and Winston. All rights reserved.

LECCIÓN **12-3**

Práctica

Usar la pendiente y la intersección

Halla la intersección con el eje x y la intersección con el eje y para cada línea. Usa las intersecciones para representar gráficamente la ecuación.

1. $x - y = -3$

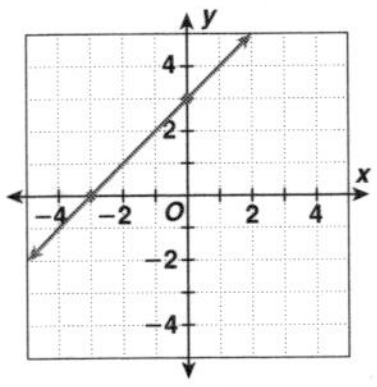

la intersección con el eje x es -3; la intersección con el y es 3

2. $2x + 3y = 12$

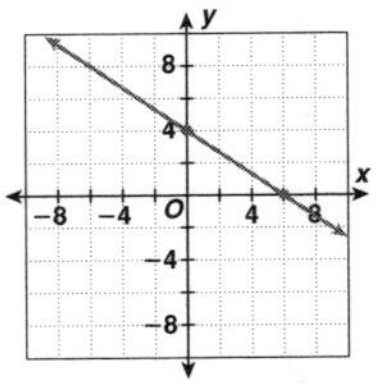

la intersección con el eje x es 6; la intersección con el eje y es 4

Escribe cada ecuación en forma de pendiente-intersección y luego halla la pendiente y la intersección con el eje y.

3. $3x + y = 0$

$y = -3x$; $m = -3$; $b = 0$

4. $2x - y = -15$

$y = 2x + 15$; $m = 2$; $b = 15$

5. $x - 5y = 10$

$y = \frac{1}{5}x - 2$; $m = \frac{1}{5}$; $b = -2$

Escribe la ecuación de la línea que pasa por cada par de puntos en forma de pendiente-intersección.

6. (3, 4), (4, 6)

$y = 2x - 2$

7. (−1, −1), (2, −10)

$y = -3x - 4$

8. (6, 5), (−9, −20)

$y = \frac{5}{3}x - 5$

9. Una pizzería cobra $8 por una pizza grande de queso, más $2 por cada sabor agregado. El precio total de una pizza grande está dado por la ecuación $C = 2t + 8$, donde t es la cantidad de sabores adicionales. Identifica la pendiente y la intersección con el eje y, y úsalas para representar gráficamente la ecuación para t entre 0 y 5 sabores.

La pendiente es 2 y la intersección con el eje y es 8.

Copyright © by Holt, Rinehart and Winston. All rights reserved.

LESSON **12-4**

Practice

Point-Slope Form

Use the point-slope form of each equation to identify a point the line passes through and the slope of the line.

1. $y - 2 = 4(x - 1)$

$m = 4$; $(x_1, y_1) = (1, 2)$

2. $y + 1 = 2(x - 3)$

$m = 2$; $(x_1, y_1) = (3, -1)$

3. $y - 4 = -3(x + 1)$

$m = -3$; $(x_1, y_1) = (-1, 4)$

4. $y + 5 = -2(x + 6)$

$m = -2$; $(x_1, y_1) = (-6, -5)$

5. $y + 4 = -9(x + 3)$

$m = -9$; $(x_1, y_1) = (-3, -4)$

6. $y - 7 = -7(x - 7)$

$m = -7$; $(x_1, y_1) = (7, 7)$

7. $y - 10 = 6(x - 8)$

$m = 6$; $(x_1, y_1) = (8, 10)$

8. $y + 12 = 2.5(x + 4)$

$m = 2.5$; $(x_1, y_1) = (-4, -12)$

9. $y + 8 = \frac{1}{2}(x - 3)$

$m = \frac{1}{2}$; $(x_1, y_1) = (3, -8)$

Write the point-slope form of the equation with the given slope that passes through the indicated point.

10. the line with slope −1 passing through (2, 5)

$y - 5 = -1(x - 2)$

11. the line with slope 2 passing through (−1, 4)

$y - 4 = 2(x + 1)$

12. the line with slope 4 passing through (−3, −2)

$y + 2 = 4(x + 3)$

13. the line with slope 3 passing through (7, −6)

$y + 6 = 3(x - 7)$

14. the line with slope −3 passing through (−6, 4)

$y - 4 = -3(x + 6)$

15. the line with slope −2 passing through (5, 1)

$y - 1 = -2(x - 5)$

16. Michael was driving at a constant speed of 60 mph when he crossed the Sandy River. After 1 hour, he passed a highway marker for mile 84. Write an equation in point-slope form, and find which highway marker he will pass 90 minutes after crossing the Sandy River.

$y - 84 = 60(x - 1)$; highway marker for 114 miles

Copyright © by Holt, Rinehart and Winston. All rights reserved.

LECCIÓN **12-4**

Práctica

Forma de punto y pendiente

Usa la forma de punto y pendiente de cada ecuación para identificar un punto por el que pasa la línea y su pendiente.

1. $y - 2 = 4(x - 1)$

$m = 4$; $(x_1, y_1) = (1, 2)$

2. $y + 1 = 2(x - 3)$

$m = 2$; $(x_1, y_1) = (3, -1)$

3. $y - 4 = -3(x + 1)$

$m = -3$; $(x_1, y_1) = (-1, 4)$

4. $y + 5 = -2(x + 6)$

$m = -2$; $(x_1, y_1) = (-6, -5)$

5. $y + 4 = -9(x + 3)$

$m = -9$; $(x_1, y_1) = (-3, -4)$

6. $y - 7 = -7(x - 7)$

$m = -7$; $(x_1, y_1) = (7, 7)$

7. $y - 10 = 6(x - 8)$

$m = 6$; $(x_1, y_1) = (8, 10)$

8. $y + 12 = 2.5(x + 4)$

$m = 2.5$; $(x_1, y_1) = (-4, -12)$

9. $y + 8 = \frac{1}{2}(x - 3)$

$m = \frac{1}{2}$; $(x_1, y_1) = (3, -8)$

Escribe la forma de punto y pendiente de la ecuación con la pendiente dada que pasa por el punto indicado.

10. La recta con pendiente −1 que pasa por (2, 5)

$y - 5 = -1(x - 2)$

11. La recta con pendiente 2 que pasa por (−1, 4)

$y - 4 = 2(x + 1)$

12. La recta con pendiente 4 que pasa por (−3, −2)

$y + 2 = 4(x + 3)$

13. La recta con pendiente 3 que pasa por (7, −6)

$y + 6 = 3(x - 7)$

14. La recta con pendiente −3 que pasa por (−6, 4)

$y - 4 = -3(x + 6)$

15. La recta con pendiente −2 que pasa por (5, 1)

$y - 1 = -2(x - 5)$

16. Michael manejaba a una velocidad constante de 60 mph cuando cruzó el río Sandy. Una hora después, pasó por la señal de la milla 84. Escribe una ecuación en forma de punto y pendiente e indica qué señal pasará a los 90 minutos de haber cruzado el río Sandy.

$y - 84 = 60(x - 1)$; la señal de las 114 millas.

Copyright © by Holt, Rinehart and Winston. All rights reserved.

Copyright © by Holt, Rinehart and Winston. All rights reserved.

LESSON **12-5**
Practice
Direct Variation

Make a graph to determine whether the data sets show direct variation.

1.

x	y
6	9
4	6
0	0
−2	−3
−8	−12

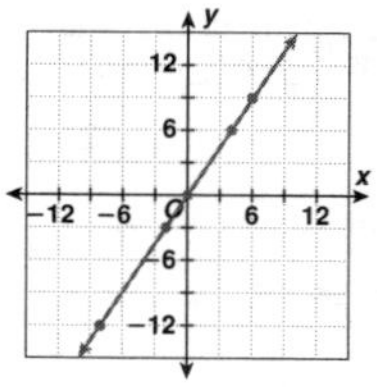

The data sets show direct variation.

2. Write the equation of direct variation for Exercise 1.

$y = 1.5x$ or $y = \frac{3}{2}x$

Find each equation of direct variation, given that *y* varies with *x*.

3. *y* is 32 when *x* is 4

$y = 8x$

4. *y* is −10 when *x* is −20

$y = \frac{1}{2}x$

5. *y* is 63 when *x* is −7

$y = -9x$

6. *y* is 40 when *x* is 50

$y = \frac{4}{5}x$

7. *y* is 87.5 when *x* is 25

$y = 3.5x$

8. *y* is 90 when *x* is 270

$y = \frac{1}{3}x$

9. The table shows the length and width of various U.S. flags. Determine whether there is direct variation between the two data sets. If so, find the equation of direct variation.

Length (ft)	2.85	5.7	7.6	9.88	11.4
Width (ft)	1.5	3	4	5.2	6

There is direct variation between the lengths and widths of the flags. $y = 1.9x$, where y is the length, x is the width, and 1.9 is the constant of proportionality

Copyright © by Holt, Rinehart and Winston. All rights reserved.

LECCIÓN **12-5**
Práctica
Variación directa

Haz una gráfica para determinar si en los conjuntos de datos se muestra una variación directa.

1.

x	y
6	9
4	6
0	0
−2	−3
−8	−12

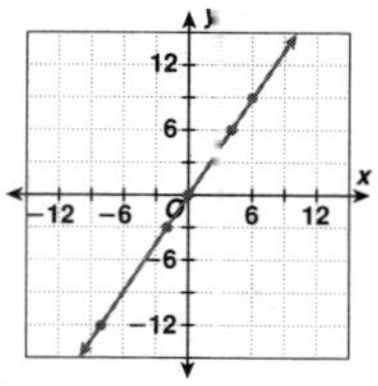

Los conjuntos de datos presentan variación directa.

2. Escribe la ecuación de variación directa para el ejercicio 1.

$y = 1.5x$ ó $y = \frac{3}{2}x$

Halla cada ecuación de variación directa, sea que *y* varía con relación a *x*.

3. *y* es 32 cuando *x* es 4

$y = 8x$

4. *y* es −10 cuando *x* es −20

$y = \frac{1}{2}x$

5. *y* es 63 cuando *x* es −7

$y = -9x$

6. *y* es 40 cuando *x* es 50

$y = \frac{4}{5}x$

7. *y* es 87.5 cuando *x* es 25

$y = 3.5x$

8. *y* es 90 cuando *x* es 270

$y = \frac{1}{3}x$

9. En la tabla se muestra la longitud y el ancho de varias banderas de Estados Unidos. Determina si hay variación directa entre los dos conjuntos de datos. Si la hay, halla la ecuación de variación directa.

Longitud (pies)	2.85	5.7	7.6	9.88	11.4
Ancho (pies)	1.5	3	4	5.2	6

Hay variación directa entre la longitud y el ancho de las banderas. $y = 1.9x$, donde y es la longitud, x es el ancho y 1.9 es la constante de proporcionalidad

Copyright © by Holt, Rinehart and Winston. All rights reserved.

LESSON **12-6**
Practice
Graphing Inequalities in Two Variables

Graph each inequality.

1. $y \geq 2x + 3$

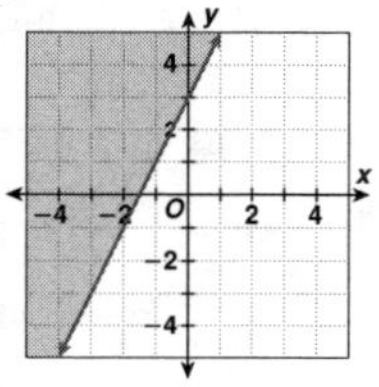

2. $y - 4x \leq 1$

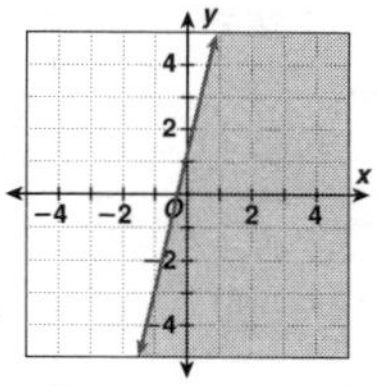

3. $2(3x - y) > 6$

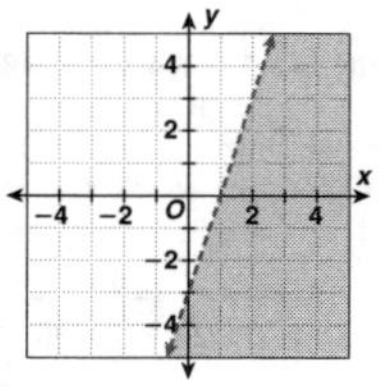

4. $y \geq \frac{3}{4}x - 1$

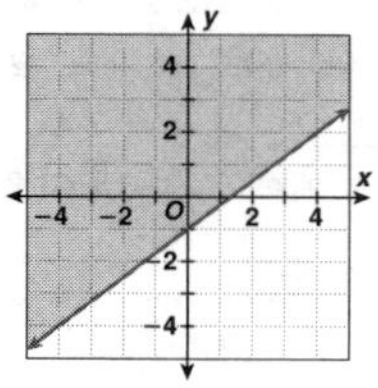

5. a. A theater club hopes to raise at least $550 on the opening night of its new show. Student tickets for the show cost $2.75, and adult tickets cost $5.50. Write and graph an inequality showing the numbers of tickets that would meet the club's goal.

$2.75x + 5.5y \geq 550$, or $y \geq -0.5x + 100$

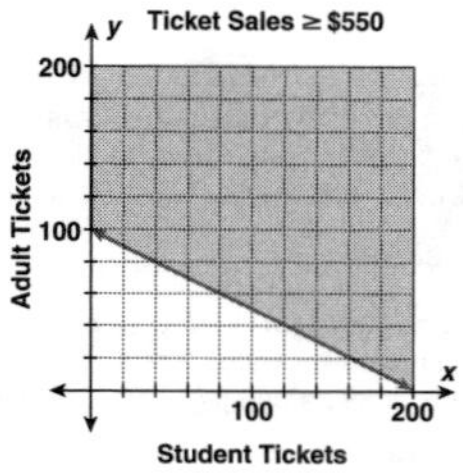

b. If the club sells 95 student tickets and 40 adult tickets, will it meet its goal?

no

Copyright © by Holt, Rinehart and Winston.

LECCIÓN **12-6**
Práctica
Representar gráficamente desigualdades de dos variables

Representa gráficamente cada desigualdad.

1. $y \geq 2x + 3$

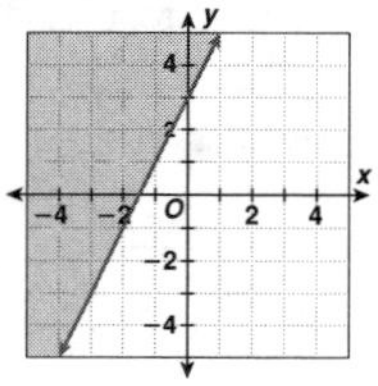

2. $y - 4x \leq 1$

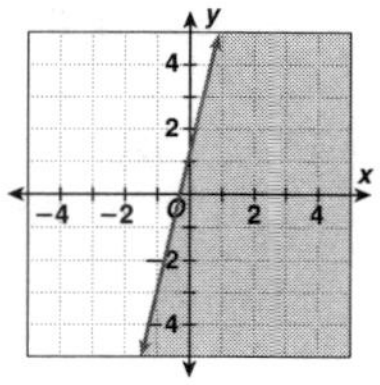

3. $2(3x - y) > 6$

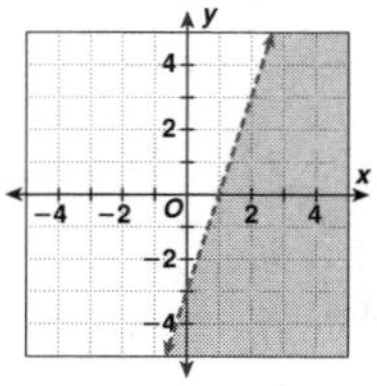

4. $y \geq \frac{3}{4}x - 1$

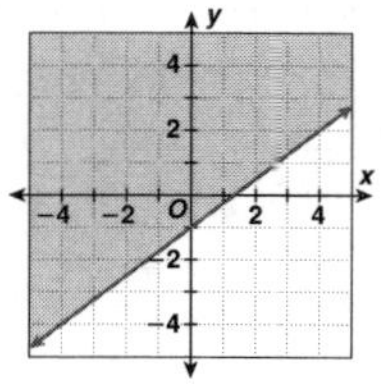

5. a. Un club de teatro espera recaudar al menos $550 en la noche de estreno de su nuevo espectáculo. Las entradas para estudiantes cuestan $2.75 y las entradas para adultos, $5.50. Escribe y representa gráficamente una desigualdad que muestre la cantidad de entradas que cumplirían el objetivo del club.

$2.75x + 5.5y \geq 550$, o $y \geq -0.5x + 100$

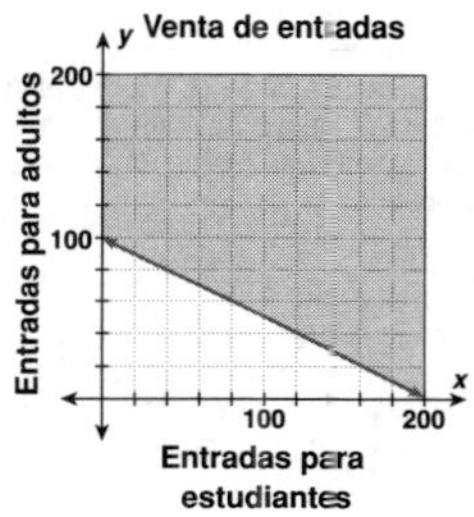

b. Si el club vende 95 entradas para estudiantes y 40 para adultos, ¿alcanzará su objetivo?

no

Copyright © by Holt, Rinehart and Winston.

Copyright © by Holt, Rinehart and Winston. All rights reserved.

LESSON **12-7**

Practice

Lines of Best Fit

Plot the data and find a line of best fit.

1.

x	20	30	50	60	80	90	110	120
y	13	20	40	54	75	82	100	112

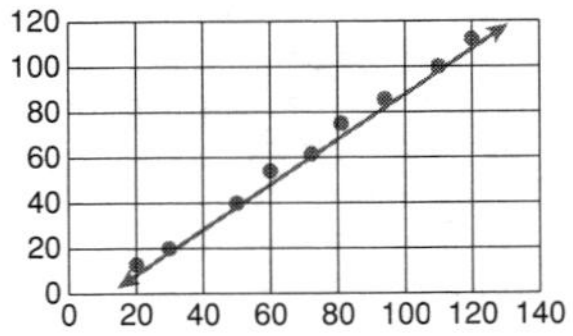

$y = x - 8$

2.

x	1.9	2.9	4.8	2.5	3.9	2.3	6.3	3.4
y	26	34	58	31	52	27	76	48

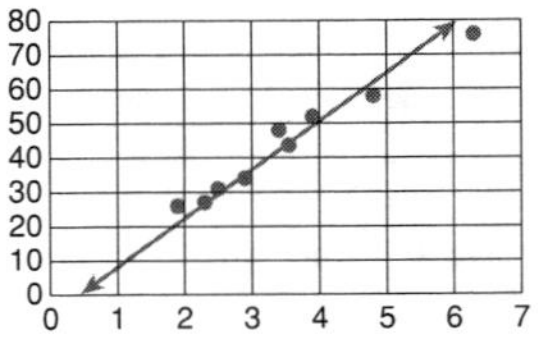

$y = 14.2x - 5.7$

3. Find the line of best fit for the student enrollment data. Use the equation of the line to predict what the enrollment at Columbus Junior High School will be in year 10. Is it reasonable to make this prediction? Explain.

Enrollment	405	485	557	593	638	712
Year	1	2	3	4	5	6

$x_m = 3.5$; $y_m = 565$; Possible equation of line of best fit: $y = 66x + 334$; Possible enrollment prediction for year 10: 994 students, which would be reasonable if the communities surrounding the school continued to grow at the same pace

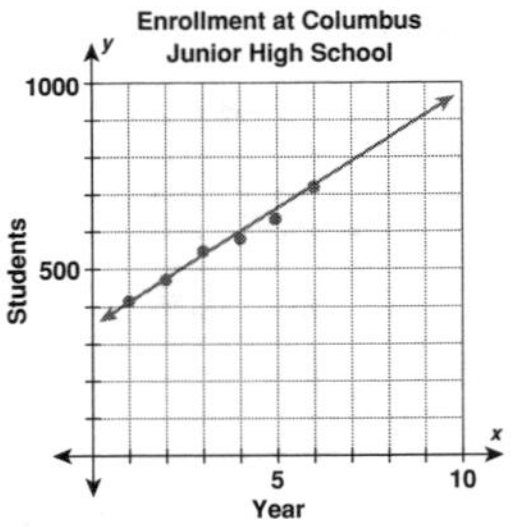

Copyright © by Holt, Rinehart and Winston. All rights reserved.

LECCIÓN **12-7**

Práctica

Líneas de mejor ajuste

Marca los datos y halla una línea de mejor ajuste.

1.

x	20	30	50	60	80	90	110	120
y	13	20	40	54	75	82	100	112

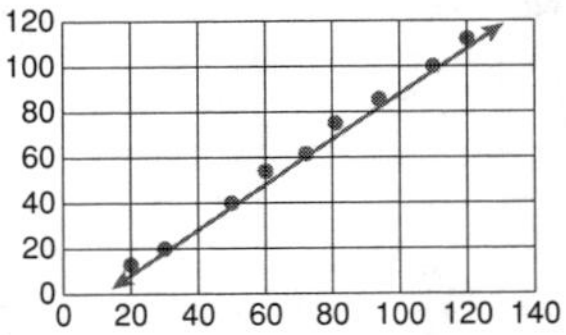

$y = x - 8$

2.

x	1.9	2.9	4.8	2.5	3.9	2.3	6.3	3.4
y	26	34	58	31	52	27	76	48

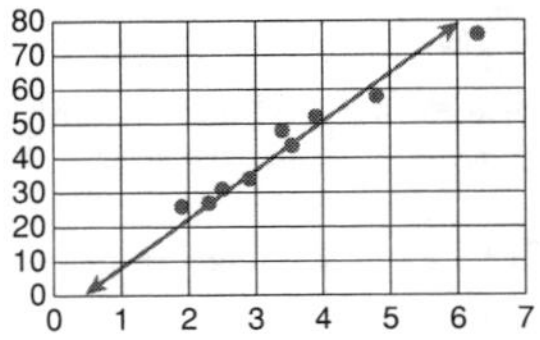

$y = 14.2x - 5.7$

3. Halla la línea de mejor ajuste para los datos de matrícula de estudiantes. Usa la ecuación de la línea para predecir cuál será la matrícula de la escuela secundaria Columbus en el año 10. ¿Es razonable hacer esta predicción? Explica.

Matrícula	405	485	557	593	638	712
Año	1	2	3	4	5	6

$x_m = 3.5$; $y_m = 565$; ecuación posible de la línea de mejor ajuste: $y = 66x + 334$; predicción de matrícula probable para el año 10: 994 estudiantes, que es un número razonable si las comunidades que rodean la escuela continúan creciendo al mismo ritmo.

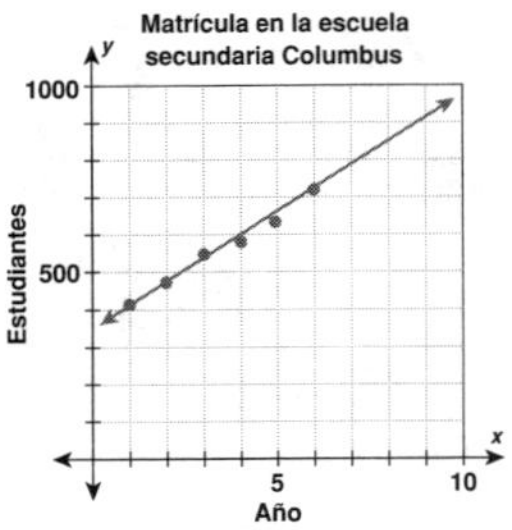

Copyright © by Holt, Rinehart and Winston. All rights reserved.

LESSON **13-1**

Practice

Terms of Arithmetic Sequences

Determine if each sequence could be arithmetic. If so, give the common difference.

1. 18, 20, 22, 24, 26, ... — 2
2. 48, 42, 36, 30, 24, ... — −6
3. 15, 30, 60, 120, 240, ... — no
4. 10.4, 8.3, 6.2, 4.1, 2, ... — −2.1
5. $\frac{1}{3}, \frac{1}{9}, \frac{1}{27}, \frac{1}{81}, \frac{1}{243}, \ldots$ — no
6. 83, 66, 49, 32, 15, ... — −17
7. 8.1, 2.7, 0.9, 0.3, 0.1, ... — no
8. $\frac{2}{3}, \frac{4}{3}, 2, \frac{8}{3}, \frac{10}{3}, \ldots$ — $\frac{2}{3}$
9. −58, −35, −12, 11, 34, ... — 23

Find the given term in each arithmetic sequence.

10. 14th term: 60, 68, 76, 84, 92, ... — 164
11. 35th term: 3.5, 3.8, 4.1, 4.4, 4.7, ... — 13.7
12. 21st term: 103, 84, 65, 46, 27, ... — −277
13. 22nd term: −2, −5, −8, −11, −14, ... — −65
14. 16th term: 73, 44, 15, −14, −43, ... — −362
15. 50th term: −9, 2, 13, 24, 35, ... — 530
16. 19th term: −87, −78, −69, −60, −51, ... — 75
17. 25th term: $3\frac{1}{4}, 3\frac{1}{2}, 3\frac{3}{4}, 4, 4\frac{1}{4}, \ldots$ — $9\frac{1}{4}$
18. A cook started with 26 ounces of special sauce. She used 1.4 ounces of the sauce in each of a number of dishes and had 2.2 ounces left over. How many dishes did she make with the sauce?

    17 dishes
19. Kuang started the basketball season with 54 points in his career. He scores 3 points more each game he plays. How many games will it take for him to have scored a total of 132 points in his basketball career?

    26 games

Copyright © by Holt, Rinehart and Winston. All rights reserved.

LECCIÓN **13-1**

Práctica

Términos de sucesiones aritméticas

Determina si cada sucesión puede ser aritmética. Si es aritmética, da la diferencia común.

1. 18, 20, 22, 24, 26, ... — 2
2. 48, 42, 36, 30, 24, ... — −6
3. 15, 30, 60, 120, 240, ... — no
4. 10.4, 8.3, 6.2, 4.1, 2, ... — −2.1
5. $\frac{1}{3}, \frac{1}{9}, \frac{1}{27}, \frac{1}{81}, \frac{1}{243}, \ldots$ — no
6. 83, 66, 49, 32, 15, ... — −17
7. 8.1, 2.7, 0.9, 0.3, 0.1, ... — no
8. $\frac{2}{3}, \frac{4}{3}, 2, \frac{8}{3}, \frac{10}{3}, \ldots$ — $\frac{2}{3}$
9. −58, −35, −12, 11, 34, ... — 23

Halla el término dado en cada sucesión aritmética.

10. 14to término: 60, 68, 76, 84, 92, ... — 164
11. 35to término: 3.5, 3.8, 4.1, 4.4, 4.7, ... — 13.7
12. 21er término: 103, 84, 65, 46, 27, ... — −277
13. 22do término: −2, −5, −8, −11, −14, ... — −65
14. 16to término: 73, 44, 15, −14, −43, ... — −362
15. 50mo término: −9, 2, 13, 24, 35, ... — 530
16. 19no término: −87, −78, −69, −60, −51, ... — 75
17. 25to término: $3\frac{1}{4}, 3\frac{1}{2}, 3\frac{3}{4}, 4, 4\frac{1}{4}, \ldots$ — $9\frac{1}{4}$
18. Una cocinera preparó 26 onzas de salsa especial. Usó 1.4 onzas de salsa en cada plato y le sobraron 2.2 onzas. ¿Cuántos platos preparó con la salsa?

    17 platos
19. Kuang comenzó la temporada de básquetbol con 54 puntos en su carrera. Marca 3 puntos más en cada partido. ¿Cuántos partidos tiene que jugar para alcanzar un total de 132 puntos?

    26 partidos

Copyright © by Holt, Rinehart and Winston. All rights reserved.

Copyright © by Holt, Rinehart and Winston. All rights reserved.

LESSON 13-2

Practice

Terms of Geometric Sequences

Determine if each sequence could be geometric. If so, give the common ratio.

1. 4, 16, 64, 256, 1024, ...   4

2. $3, \frac{3}{2}, \frac{3}{4}, \frac{3}{8}, \frac{3}{16}, \ldots$ $\frac{1}{2}$

3. 5, 10, 15, 20, 25, ...   not

4. 3, 18, 108, 648, 3888, ...   6

5. 1250, 125, 12.5, 1.25, 0.125, ...   0.1

6. 10, 15, 22.5, 33.75, 50.625, ...   1.5

7. $36, 12, 4, \frac{4}{3}, \frac{4}{9}, \ldots$ $\frac{1}{3}$

8. 1440, 720, 240, 60, 12, ...   not

9. 9, 3, 1, 0.5, 0.25, ...   not

Find the given term in each geometric sequence.

10. 6th term: 25, 75, 225, 675, ...   6075

11. 10th term: 320, 160, 80, 40, ...   0.625

12. 9th term: 4.5, 9, 18, 36, ...   1152

13. 7th term: 0.02, 0.2, 2, 20, ...   20,000

14. 12th term: $\frac{1}{1000}, \frac{1}{100}, \frac{1}{10}, 1, \ldots$ 100,000,000

15. 8th term: $\frac{3}{8}, \frac{3}{4}, \frac{3}{2}, 3, \ldots$ 48

16. In an experiment a population of flies triples every week. The experiment starts with 12 flies. How many flies will there be by the end of week 5?

2916 flies

17. A small business earned $21 in its first month. It quadrupled this amount each month for the next several months. How much did the business earn in the 4th month?

$1344

Copyright © by Holt, Rinehart and Winston. All rights reserved.

LECCIÓN 13-2

Práctica

Términos de sucesiones geométricas

Determina si cada sucesión puede ser geométrica. Si es geométrica, da la razón común.

1. 4, 16, 64, 256, 1024, ...   4

2. $3, \frac{3}{2}, \frac{3}{4}, \frac{3}{8}, \frac{3}{16}, \ldots$ $\frac{1}{2}$

3. 5, 10, 15, 20, 25, ...   no

4. 3, 18, 108, 648, 3888, ...   6

5. 1250, 125, 12.5, 1.25, 0.125, ...   0.1

6. 10, 15, 22.5, 33.75, 50.625 ...   1.5

7. $36, 12, 4, \frac{4}{3}, \frac{4}{9}, \ldots$ $\frac{1}{3}$

8. 1440, 720, 240, 60, 12, ...   no

9. 9, 3, 1, 0.5, 0.25, ...   no

Halla el término dado en cada sucesión geométrica.

10. 6to término: 25, 75, 225, 675, ...   6075

11. 10mo término: 320, 160, 80, 40, ...   0.625

12. 9no término: 4.5, 9, 18, 36, ...   1152

13. 7mo término: 0.02, 0.2, 2, 20, ...   20,000

14. 12do término: $\frac{1}{1000}, \frac{1}{100}, \frac{1}{10}, 1, \ldots$ 100,000,000

15. 8vo término: $\frac{3}{8}, \frac{3}{4}, \frac{3}{2}, 3, \ldots$ 48

16. En un experimento, la población de moscas se triplica cada semana. El experimento comienza con 12 moscas. ¿Cuántas habrá al final de la quinta semana?

2916 moscas

17. Un pequeño comercio ganó $21 el primer mes. Cuadruplicó esta cantidad durante varios de los meses siguientes. ¿Cuánto ganó el comercio en el cuarto mes?

$1344

Copyright © by Holt, Rinehart and Winston. All rights reserved.

LESSON 13-3

Practice

Other Sequences

Use first and second differences to find the next three terms in each sequence.

1. 3, 6, 10, 15, 21, ...   28, 36, 45

2. 11, 14, 18, 25, 37, ...   56, 84, 123

3. $10, 16, 22\frac{1}{3}, 29, 36, \ldots$ $43\frac{1}{3}$, 51, 59

4. 14.5, 22.5, 31, 40, 49.5, ...   59.5, 70, 81

Give the next three terms in each sequence using the simplest rule you can find.

5. 6, 7, 10, 19, 38, ...   71, 122, 195

6. 0.5, 2, 4.5, 8, 12.5, ...   18, 24.5, 32

7. 36, 55, 80, 111, 148, ...   191, 240, 295

8. 3, 10, 21, 36, 55, ...   78, 105, 136

9. 1, 6, 15, 28, 45, ...   66, 91, 120

10. 0, 11, 30, 57, 92, ...   135, 186, 245

Find the first five terms of each sequence defined by the given rule.

11. $a_n = \frac{n^2+2}{n}$ $3, 3, 3\frac{2}{3}, 4\frac{1}{2}, 5\frac{2}{5}$

12. $a_n = \frac{5n-2}{n+1}$ $1\frac{1}{2}, 2\frac{2}{3}, 3\frac{1}{4}, 3\frac{3}{5}, 3\frac{5}{6}$

13. $a_n = \frac{3n^2}{n+2}$ $1, 3, 5\frac{2}{5}, 8, 10\frac{5}{7}$

14. Suppose *a*, *b*, and *c* are three consecutive numbers in the Fibonacci sequence. Complete the following table and guess the pattern.

a, b, c	*ab*	*bc*
1, 1, 2	1	2
2, 3, 5	6	15
5, 8, 13	40	104
13, 21, 34	273	714
34, 55, 89	1870	4895

The difference of *bc* and *ab* is the square of *b*.

Copyright © by Holt, Rinehart and Winston. All rights reserved.

LECCIÓN 13-3

Práctica

Otras sucesiones

Usa las primeras y las segundas diferencias para hallar los tres términos siguientes en cada sucesión.

1. 3, 6, 10, 15, 21, ...   28, 36, 45

2. 11, 14, 18, 25, 37, ...   56, 84, 123

3. $10, 16, 22\frac{1}{3}, 29, 36, \ldots$ $43\frac{1}{3}$, 51, 59

4. 14.5, 22.5, 31, 40, 49.5, ...   59.5, 70, 81

Halla los tres términos siguientes de la sucesión usando la regla más simple que puedas hallar.

5. 6, 7, 10, 19, 38, ...   71, 122, 195

6. 0.5, 2, 4.5, 8, 12.5, ...   18, 24.5, 32

7. 36, 55, 80, 111, 148, ...   191, 240, 295

8. 3, 10, 21, 36, 55, ...   78, 105, 136

9. 1, 6, 15, 28, 45, ...   66, 91, 120

10. 0, 11, 30, 57, 92, ...   135, 186, 245

Halla los primeros cinco términos de cada sucesión definida por la regla dada.

11. $a_n = \frac{n^2+2}{n}$ $3, 3, 3\frac{2}{3}, 4\frac{1}{2}, 5\frac{2}{5}$

12. $a_n = \frac{5n-2}{n+1}$ $1\frac{1}{2}, 2\frac{2}{3}, 3\frac{1}{4}, 3\frac{3}{5}, 3\frac{5}{6}$

13. $a_n = \frac{3r^2}{n+2}$ $1, 3, 5\frac{2}{5}, 8, 10\frac{5}{7}$

14. Supongamos que *a*, *b*, y *c* son tres números consecutivos en la sucesión de Fibonacci. Completa la siguiente tabla y adivina el patrón.

a, b, c	*ab*	*bc*
1, 1, 2	1	2
2, 3, 5	6	15
5, 8, 13	40	104
13, 21, 34	273	714
34, 55, 89	1870	4895

La diferencia entre *bc* y *ab* es el cuadrado de *b*.

Copyright © by Holt, Rinehart and Winston. All rights reserved.

Copyright © by Holt, Rinehart and Winston. All rights reserved.

LESSON **13-4**
Practice
Linear Functions

Determine whether each function is linear.

1. $f(x) = -3x + 2$

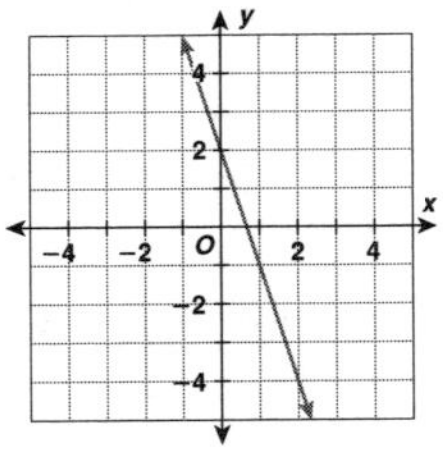

linear

2. $f(x) = x^2 - 1$

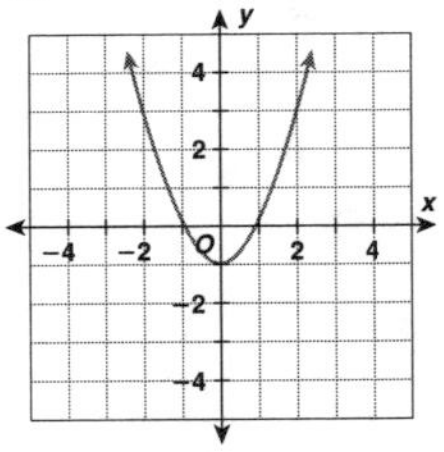

not linear

Write a rule for each linear function.

3.

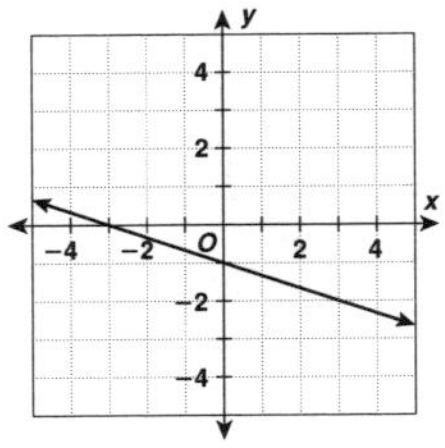

$f(x) = -\frac{1}{3}x - 1$

4.

x	y
−3	16
−1	12
3	4
7	−4

$f(x) = -2x + 10$

5. At the Sweater Store, the price of a sweater is 20% more than the wholesale cost, plus a markup of $8. Find a rule for a linear function that describes the price of sweaters at the Sweater Store. Use it to determine the price of a sweater with a wholesale cost of $24.50.

$f(x) = 1.2x + 8$, where x is the wholesale cost of the sweater.

$f(24.50) = 1.2 \bullet 24.5 + 8 = \$37.40$

Copyright © by Holt, Rinehart and Winston. All rights reserved.

LECCIÓN **13-4**
Práctica
Funciones lineales

Determina si cada función es lineal.

1. $f(x) = -3x + 2$

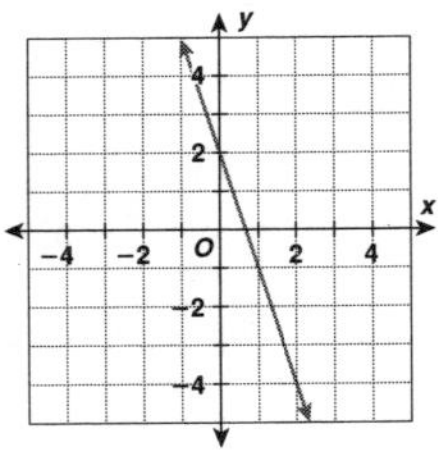

lineal

2. $f(x) = x^2 - 1$

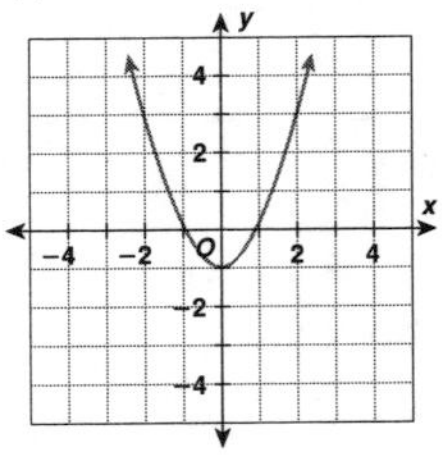

no lineal

Escribe una regla para cada función lineal.

3.

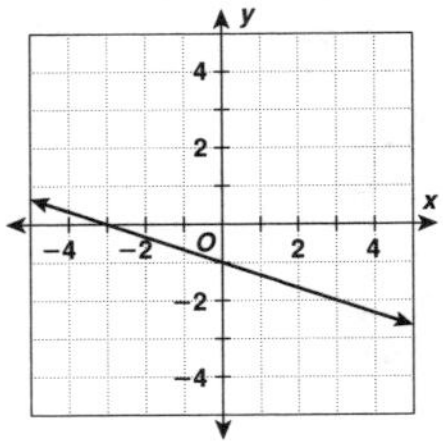

$f(x) = -\frac{1}{3}x - 1$

4.

x	y
−3	16
−1	12
3	4
7	−4

$f(x) = -2x + 10$

5. En la tienda El Suéter, cada suéter cuesta un 20% más que el precio al por mayor más un margen de $8. Halla una regla para una función lineal que dé cuenta del precio de los suéteres de la tienda. Usa la regla para determinar el precio de un suéter con un precio al por mayor de $24.50.

$f(x) = 1.2x + 8$, donde x es el costo al por mayor del suéter.

$f(24.50) = 1.2 \bullet 24.5 + 8 = \$37.40$

Copyright © by Holt, Rinehart and Winston. All rights reserved.

LESSON **13-5**
Practice
Exponential Functions

Create a table for each exponential function, and use it to graph the function.

1. $f(x) = 0.5 \cdot 4^x$

x	y
−1	$y = 0.5 \cdot 4^{-1} = 0.125$
0	$y = 0.5 \bullet 4^0 = 0.5$
1	$y = 0.5 \bullet 4^1 = 2$
2	$y = 0.5 \bullet 4^2 = 8$

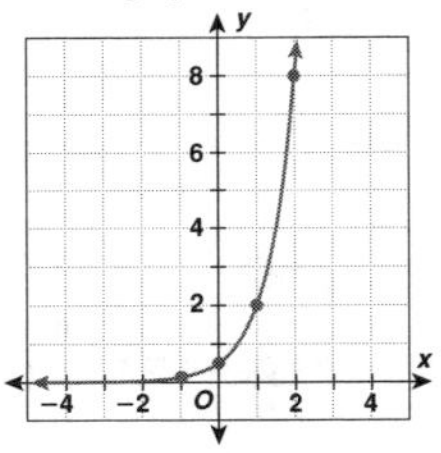

2. $f(x) = \frac{1}{3} \cdot 3^x$

x	y
−1	$y = \frac{1}{3} \cdot 3^{-1} = \frac{1}{9}$
0	$y = \frac{1}{3} \bullet 3^0 = \frac{1}{3}$
1	$y = \frac{1}{3} \bullet 3^1 = 1$
2	$y = \frac{1}{3} \bullet 3^2 = 3$

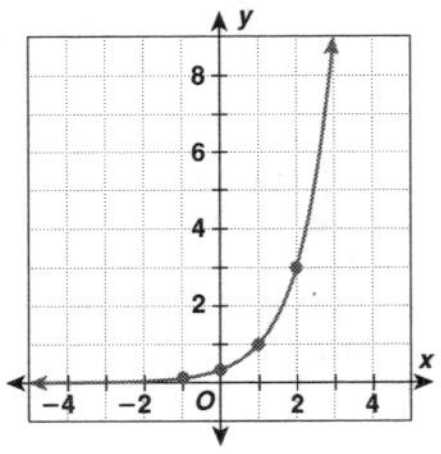

3. A forestry department introduce 500 fish to a lake. The fish are expected to increase at a rate of 35% each year. Write an exponential function to calculate the number of fish in the lake at the end of each year. Predict how many fish will be in the lake at the end of 5 years.

$f(x) = 500\,(1.35)^x$; 2242 fish

4. A stock valued at $756 has been declining steadily at the rate of 4% a year for the last few years. If this decline continues, predict what the value of the stock will be at the end of 3 years.

$668.86

5. Todd's starting salary at his new job is $400 a week. He is promised a 3% increase in salary every year. Predict to the nearest dollar what Todd's expected yearly salary will be after working for 4 years.

$23,411

Copyright © by Holt, Rinehart and Winston. All rights reserved.

LECCIÓN **13-5**
Práctica
Funciones exponenciales

Crea una tabla para cada función exponencial y úsala para representar gráficamente la función.

1. $f(x) = 0.5 \cdot 4^x$

x	y
−1	$y = 0.5 \cdot 4^{-1} = 0.125$
0	$y = 0.5 \bullet 4^0 = 0.5$
1	$y = 0.5 \bullet 4^1 = 2$
2	$y = 0.5 \bullet 4^2 = 8$

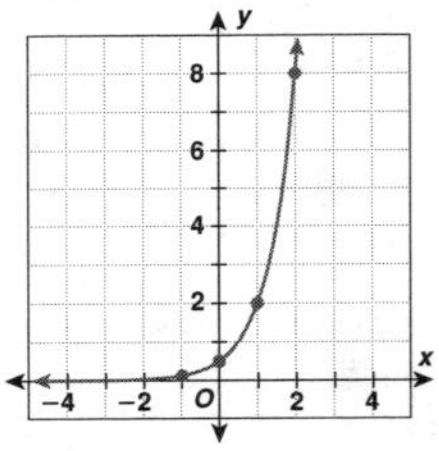

2. $f(x) = \frac{1}{3} \cdot 3^x$

x	y
−1	$y = \frac{1}{3} \cdot 3^{-1} = \frac{1}{9}$
0	$y = \frac{1}{3} \bullet 3^0 = \frac{1}{3}$
1	$y = \frac{1}{3} \bullet 3^1 = 1$
2	$y = \frac{1}{3} \bullet 3^2 = 3$

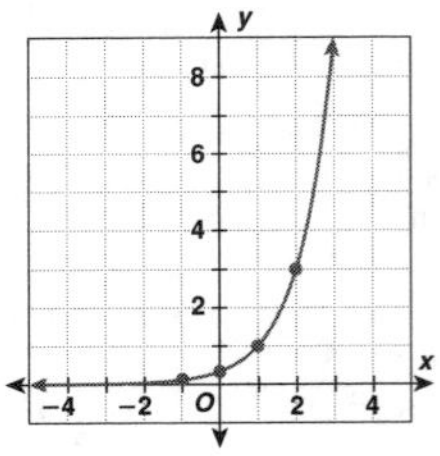

3. Un departamento de ingeniería forestal introduce 500 peces en un lago. Se espera que la población de peces crezca a una tasa de 35% por año. Escribe una función exponencial para calcular la cantidad de peces que habrá al final de cada año. Predice cuántos peces habrá en el lago al cabo de 5 años.

$f(x) = 500\,(1.35)^x$; 2242 peces

4. El valor de ciertos bonos valuados en $756 se ha reducido a una tasa constante de 4% por año durante los últimos años. Si esta reducción continúa, predice cuál será el valor de los bonos al cabo de tres años.

$668.86

5. El salario inicial de Todd en su nuevo trabajo es de $400 por semana. Le han prometido un aumento de 3% de su salario por año. Predice, redondeando al dólar más cercano, cuál será el salario anual de Todd al cabo de 4 años.

$23,411

Copyright © by Holt, Rinehart and Winston. All rights reserved.

Copyright © by Holt, Rinehart and Winston. All rights reserved.

LESSON 13-6
Practice
Quadratic Functions

Create a table for each quadratic function, and use it to make a graph.

1. $f(x) = x^2 - 5$

x	$f(x) = x^2 - 5$
−3	$f(-3) = (-3)^2 - 5 = 4$
−1	$f(-1) = (-1)^2 - 5 = -4$
0	$f(0) = (0)^2 - 5 = -5$
2	$f(2) = (2)^2 - 5 = -1$
3	$f(3) = (3)^2 - 5 = 4$

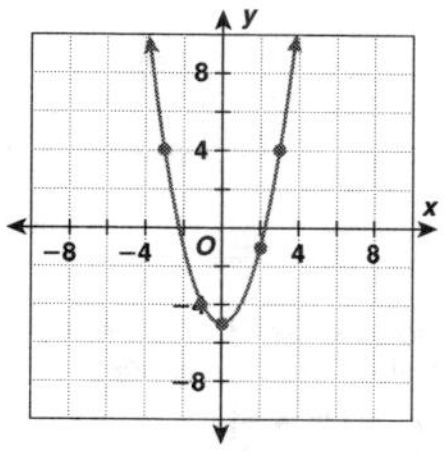

2. $f(x) = x^2 - 2x + 3$

x	$f(x) = x^2 - 2x + 3$
3	$f(3) = (3)^2 - 2(3) + 3 = 6$
2	$f(2) = (2)^2 - 2(2) + 3 = 3$
1	$f(1) = (1)^2 - 2(1) + 3 = 2$
0	$f(0) = (0)^2 - 2(0) + 3 = 3$
−1	$f(-1) = (-1)^2 - 2(-1) + 3 = 6$

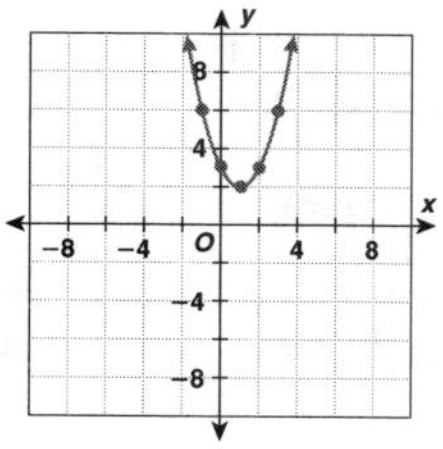

3. Find $f(-3)$, $f(0)$, $f(3)$ for each quadratic function.

	$f(-3)$	$f(0)$	$f(3)$
$f(x) = x^2 - 2x + 1$	16	1	4
$f(x) = x^2 - 6$	3	−6	3
$f(x) = x^2 - x + 3$	15	3	9

4. The function $f(t) = -4.9t^2$ gives the distance in meters that an object will fall toward Earth in t seconds. Find the distance an object will fall in 1, 2, 3, 4, and 5 seconds. (Note that the distance traveled by a falling object is shown by a negative number.)

4.9 m, 19.6 m, 44.1 m, 78.4 m, and 122.5 m

Copyright © by Holt, Rinehart and Winston. All rights reserved.

LECCIÓN 13-6
Práctica
Funciones cuadráticas

Crea una tabla para cada función cuadrática y úsala para hacer una gráfica.

1. $f(x) = x^2 - 5$

x	$f(x) = x^2 - 5$
−3	$f(-3) = (-3)^2 - 5 = 4$
−1	$f(-1) = (-1)^2 - 5 = -4$
0	$f(0) = (0)^2 - 5 = -5$
2	$f(2) = (2)^2 - 5 = -1$
3	$f(3) = (3)^2 - 5 = 4$

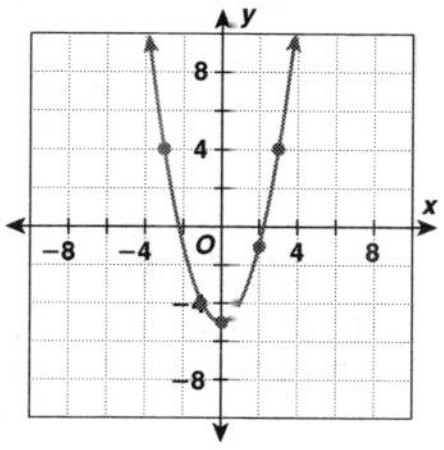

2. $f(x) = x^2 - 2x + 3$

x	$f(x) = x^2 - 2x + 3$
3	$f(3) = (3)^2 - 2(3) + 3 = 6$
2	$f(2) = (2)^2 - 2(2) + 3 = 3$
1	$f(1) = (1)^2 - 2(1) + 3 = 2$
0	$f(0) = (0)^2 - 2(0) + 3 = 3$
−1	$f(-1) = (-1)^2 - 2(-1) + 3 = 6$

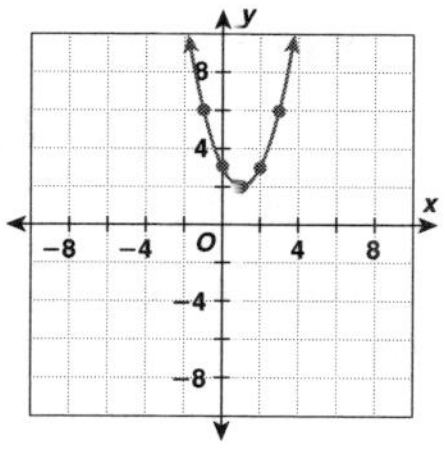

3. Halla $f(-3)$, $f(0)$, $f(3)$ para cada función cuadrática.

	$f(-3)$	$f(0)$	$f(3)$
$f(x) = x^2 - 2x + 1$	16	1	4
$f(x) = x^2 - 6$	3	−6	3
$f(x) = x^2 - x + 3$	15	3	9

4. La función $f(t) = -4.9t^2$ indica la distancia en metros que recorre un objeto cuando cae hacia la Tierra en t segundos. Halla la distancia que recorre un objeto en 1, 2, 3, 4 y 5 segundos. (Toma en cuenta que la distancia que recorre un objeto que cae se representa con un número negativo).

4.9 m, 19.6 m, 44.1 m, 78.4 m, y 122.5 m

Copyright © by Holt, Rinehart and Winston. All rights reserved.

LESSON 13-7
Practice
Inverse Variation

Tell whether each relationship is an inverse variation.

1. The table shows the length and width of certain rectangles.

Length	6	8	12	16	24
Width	8	6	4	3	2

The relationship is an inverse variation.

2. The table shows the number of days needed to paint a house for the size of the work crew.

Crew Size	2	3	4	5	6
Days of Painting	21	14	10.5	8.5	7

The relationship is not an inverse variation.

3. The table shows the time spent traveling at different speeds.

Hours	5	6	8	9	12
mi/h	72	60	45	40	30

The relationship is an inverse variation.

Graph each inverse variation function.

4. $f(x) = \frac{4}{x}$

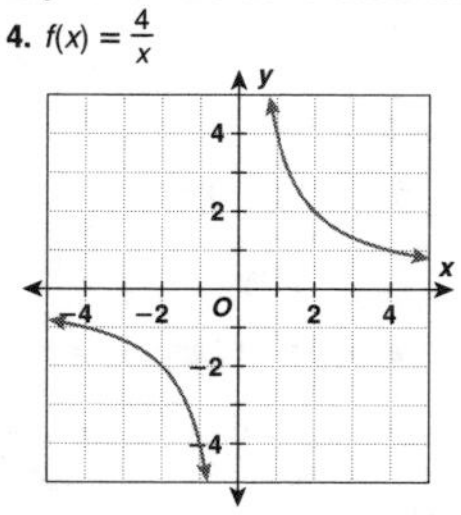

5. $f(x) = \frac{5}{x}$

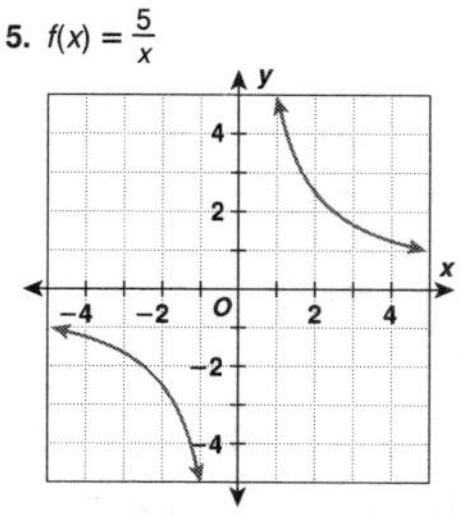

6. Amperes (abbreviated amp) measure the strength of electric current. An ohm is the unit of electrical resistance. In an electric circuit, the current varies inversely as the resistance. If the current is 24 amps when the resistance is 20 ohms, find the inverse variation function and use it to find the resistance in ohms when the current is 40 amps.

$y = \frac{480}{x}$; 12 ohms

Copyright © by Holt, Rinehart and Winston. All rights reserved.

LECCIÓN 13-7
Práctica
Variación inversa

Indica si cada relación es una variación inversa.

1. En la tabla se muestra la longitud y el ancho de ciertos rectángulos.

Longitud	6	8	12	16	24
Ancho	8	6	4	3	2

La relación es una variación inversa.

2. En la tabla se muestra la cantidad de días necesarios para pintar una casa según la cantidad de trabajadores.

Cantidad de trabajadores	2	3	4	5	6
Días de trabajo	21	14	10.5	8.5	7

La relación no es una variación inversa.

3. En la tabla se muestra el tiempo que una persona viajó a distintas velocidades.

Horas	5	6	8	9	12
mph	72	60	45	40	30

La relación es una variación inversa.

Representa gráficamente cada función de variación inversa.

4. $f(x) = \frac{4}{x}$

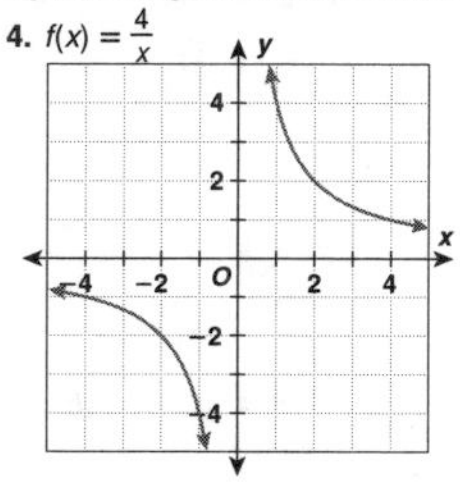

5. $f(x) = \frac{5}{x}$

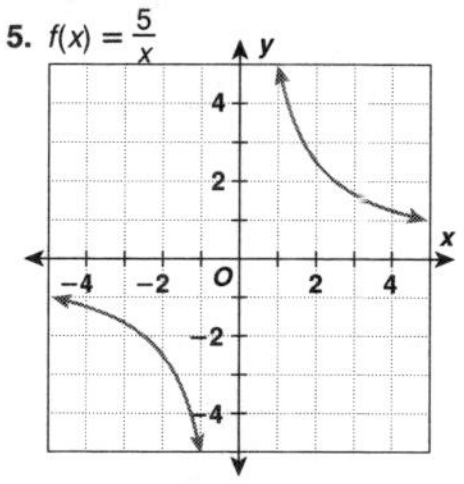

6. Los amperios (se abrevia amp) miden la intensidad de corriente eléctrica. El ohmio es la unidad de resistencia eléctrica. En un circuito eléctrico, la corriente varía inversamente con relación a la resistencia. Si la corriente es de 24 amp cuando la resistencia es de 20 ohmios, halla la función de variación inversa y úsala para hallar la resistencia en ohmios cuando la corriente es de 40 amp.

$y = \frac{480}{x}$; 12 ohmios

Copyright © by Holt, Rinehart and Winston. All rights reserved.

Copyright © by Holt, Rinehart and Winston. All rights reserved.

LESSON **14-1**
Practice
Polynomials

Determine whether each expression is a monomial.

1. $-135x^5$ — yes
2. $2.4x^3y^{19}$ — yes
3. $\frac{2p^2}{q^3}$ — no
4. $3r^{\frac{1}{2}}$ — no
5. $43a^2b^{6.1}$ — no
6. $\frac{7}{9}x^2yz^5$ — yes

Classify each expression as a monomial, a binomial, a trinomial, or not a polynomial.

7. $-8.9xy + \frac{6}{y^5}$ — not a polynomial
8. $\frac{9}{8}ab^8c^2d$ — monomial
9. $x^8 + x + 1$ — trinomial
10. $-7pq^{-2}r^4$ — not a polynomial
11. $5n^{15} - 9n + \frac{1}{3}$ — trinomial
12. $r^8 - 5.5r^{75}$ — binomial

Find the degree of each polynomial.

13. $7 - 14x$ — 1
14. $5a + a^2 + \frac{6}{7}a^3$ — 3
15. $7w - 16u + 3v$ — 1
16. $9p - 9q - 9p^3 - 9q^2$ — 3
17. $z^9 + 10y^8 - x$ — 9
18. $100{,}050 + \frac{4}{5}k - k^4$ — 4

19. The volume of a box with height x, length $x - 1$, and width $2x + 2$ is given by the trinomial $2x^3 - 2x$. What is the volume of the box if its height is 4 feet?

$120\ \text{ft}^3$

20. The trinomial $-16t^2 + 32t + 32$ describes the height in feet of a ball thrown upward after t seconds. What is the height of the ball $\frac{5}{8}$ seconds after it was thrown?

45.75 feet

Copyright © by Holt, Rinehart and Winston.
All rights reserved.

LECCIÓN **14-1**
Práctica
Polinomios

Determina si las siguientes expresiones son monomios.

1. $-135x^5$ — sí
2. $2.4x^3y^{19}$ — sí
3. $\frac{2p^2}{q^3}$ — no
4. $3r^{\frac{1}{2}}$ — no
5. $43a^2b^{6.1}$ — no
6. $\frac{7}{9}x^2yz^5$ — sí

Indica si las expresiones son monomios, binomios, trinomios o si no son polinomios.

7. $-8.9xy + \frac{6}{y^5}$ — no es un polinomio
8. $\frac{9}{8}ab^8c^2d$ — monomio
9. $x^8 + x + 1$ — trinomio
10. $-7pq^{-2}r^4$ — no es un polinomio
11. $5n^{15} - 9n + \frac{1}{3}$ — trinomio
12. $r^8 - 5.5r^{75}$ — binomio

Halla el grado de cada polinomio.

13. $7 - 14x$ — 1
14. $5a + a^2 + \frac{6}{7}a^3$ — 3
15. $7w - 16u + 3v$ — 1
16. $9p - 9q - 9p^3 - 9q^2$ — 3
17. $z^9 + 10y^8 - x$ — 9
18. $100{,}050 + \frac{4}{5}k - k^4$ — 4

19. El volumen de una caja de altura x, longitud $x - 1$, y ancho $2x + 2$ está dado por el trinomio $2x^3 - 2x$. ¿Cuál es el volumen de la caja si mide 4 pies de altura?

$120\ \text{pies}^3$

20. El trinomio $-16t^2 + 32t + 32$ describe la altura en pies de una pelota que se lanza hacia arriba luego de t segundos. ¿Qué altura alcanza la pelota pasados los $\frac{5}{8}$ de segundos desde que fue lanzada?

45.75 pies

Copyright © by Holt, Rinehart and Winston.
All rights reserved.

LESSON **14-2**
Practice
Simplifying Polynomials

Identify the like terms in each polynomial.

1. $x^2 - 8x + 3x^2 + 6x - 1$ — x^2 and $3x^2$, $-8x$ and $6x$
2. $2c^2 + d^3 + 3d^3 - 2c^2 + 6$ — $2c^2$ and $-2c^2$, d^3 and $3d^3$
3. $2x^2 - 2xy - 2y^2 + 3xy + 3x^2$ — $2x^2$ and $3x^2$, $-2xy$ and $3xy$
4. $2 - 9x + x^2 - 3 + x$ — $-9x$ and x, 2 and -3
5. $xy - 5x + y - x + 10y - 3y^2$ — $-5x$ and $-x$, y and $10y$
6. $6p + 2p^2 + pq + 2q^3 - 2p$ — $6p$ and $-2p$
7. $3a + 2b + a^2 - 5b + 7a$ — $3a$ and $7a$, $2b$ and $-5b$
8. $10m - 3m^2 + 9m^2 - 3m - m^3$ — $10m$ and $-3m$, $-3m^2$ and $9m^2$

Simplify.

9. $2h - 9hk + 6h - 6k$ — $8h - 9hk - 6k$
10. $9(x^2 + 2xy - y^2) - 2(x^2 + xy)$ — $7x^2 + 16xy - 9y^2$
11. $7qr - q^2r^3 + 2q^2r^3 - 6qr$ — $q^2r^3 + qr$
12. $8v^4 + 3v^2 + 2v^2 - 16$ — $8v^4 + 5v^2 - 16$
13. $3(x + 2y) + 2(2x - 3y)$ — $7x$
14. $7(1 - x) + 3x^2y + 7x - 7$ — $3x^2y$
15. $6(9y + 1) + 8(2 - 3y)$ — $30y + 22$
16. $a^2b - a^2 + ab^2 - 3a^2b + ab$ — $-2a^2b - a^2 + ab^2 + ab$

17. A student in Tracey's class created the following expression: $y^3 - 3y + 4(y^2 - y^3)$. Use the Distributive Property to write an equivalent expression.

$-3y^3 + 4y^2 - 3y$

Copyright © by Holt, Rinehart and Winston.
All rights reserved.

LECCIÓN **14-2**
Práctica
Cómo simplificar polinomios

Identifica los términos semejantes en cada polinomio.

1. $x^2 - 8x + 3x^2 + 6x - 1$ — x^2 y $3x^2$, $-8x$ y $6x$
2. $2c^2 + d^3 + 3d^3 - 2c^2 + 6$ — $2c^2$ y $-2c^2$, d^3 y $3d^3$
3. $2x^2 - 2xy - 2y^2 + 3xy + 3x^2$ — $2x^2$ y $3x^2$, $-2xy$ y $3xy$
4. $2 - 9x + x^2 - 3 + x$ — $-9x$ y x, 2 y -3
5. $xy - 5x + y - x + 10y - 3y^2$ — $-5x$ y $-x$, y y $10y$
6. $6p + 2p^2 + pq + 2q^3 - 2p$ — $6p$ y $-2p$
7. $3a + 2b + a^2 - 5b + 7a$ — $3a$ y $7a$, $2b$ y $-5b$
8. $10m - 3m^2 + 9m^2 - 3m - m^3$ — $10m$ y $-3m$, $-3m^2$ y $9m^2$

Simplifica.

9. $2h - 9hk + 6h - 6k$ — $8h - 9hk - 6k$
10. $9(x^2 + 2xy - y^2) - 2(x^2 + xy)$ — $7x^2 + 16xy - 9y^2$
11. $7qr - q^2r^3 + 2q^2r^3 - 6qr$ — $q^2r^3 + qr$
12. $8v^4 + 3v^2 + 2v^2 - 16$ — $8v^4 + 5v^2 - 16$
13. $3(x + 2y) + 2(2x - 3y)$ — $7x$
14. $7(1 - x) + 3x^2y + 7x - 7$ — $3x^2y$
15. $6(9y + 1) + 8(2 - 3y)$ — $30y + 22$
16. $a^2b - a^2 + ab^2 - 3a^2b + ab$ — $-2a^2b - a^2 + ab^2 + ab$

17. Un estudiante de la clase de Tracey propuso la siguiente expresión: $y^3 - 3y + 4(y^2 - y^3)$. Usa la propiedad distributiva para escribir una expresión equivalente.

$-3y^3 + 4y^2 - 3y$

Copyright © by Holt, Rinehart and Winston.
All rights reserved.

Copyright © by Holt, Rinehart and Winston.
All rights reserved.

LESSON 14-3

Practice

Adding Polynomials

Add.

1. $(a^2 + a + 3) + (15a^2 + 2a + 9)$

 $16a^2 + 3a + 12$

2. $(5x + 2x^2) + (3x - 2x^2)$

 $8x$

3. $(mn - 10 + mn^2) + (5 + 3mn - 4mn^2)$

 $-3mn^2 + 4mn - 5$

4. $(7y^2z + 9 + yz^2) + (y^2z - 2yz^2)$

 $8y^2z - yz^2 + 9$

5. $(s^3 + 3s - 3) + (2s^3 + 9s - 2) + (s - s^3)$

 $2s^3 + 13s - 5$

6. $(6wv - 4w^2v + 7wv^2) + (5w^2v - 7wv^2) + (wv^2 - 5wv + 6w^2v)$

 $7w^2v + wv^2 + wv$

7. $(6b^2c^2 - 4b^2c + 3bc) + (9b^2c^2 - 4bc + 12) + (2b^2c - 3bc - 8)$

 $15b^2c^2 - 2b^2c - 4bc + 4$

8. $(7e^2 + 3e + 2) + (9 - 6e + 4e^2) + (9e + 2 - 6e^2) + (4e^2 - 7e + 8)$

 $9e^2 - e + 21$

9. $(f^4g - fg^3 + 2fg - 4) + (3fg^3 + 3) + (4f^4g - 5fg) + (3 - 12fg^3 + f^4g)$

 $6f^4g - 10fg^3 - 3fg + 2$

10. Six blocks of height $4h + 4$ each and 3 blocks of height $8 - 2h$ each are stacked on top of each other to form one big tower. Find an expression for the overall height of the tower.

 $18h + 48$

Copyright © by Holt, Rinehart and Winston. All rights reserved.

LECCIÓN 14-3

Práctica

Cómo sumar polinomios

Suma.

1. $(a^2 + a + 3) + (15a^2 + 2a + 9)$

 $16a^2 + 3a + 12$

2. $(5x + 2x^2) + (3x - 2x^2)$

 $8x$

3. $(mn - 10 + mn^2) + (5 + 3mn - 4mn^2)$

 $-3mn^2 + 4mn - 5$

4. $(7y^2z + 9 + yz^2) + (y^2z - 2yz^2)$

 $8y^2z - yz^2 + 9$

5. $(s^3 + 3s - 3) + (2s^3 + 9s - 2) + (s - s^3)$

 $2s^3 + 13s - 5$

6. $(6wv - 4w^2v + 7wv^2) + (5w^2v - 7wv^2) + (wv^2 - 5wv + 6w^2v)$

 $7w^2v + wv^2 + wv$

7. $(6b^2c^2 - 4b^2c + 3bc) + (9b^2c^2 - 4bc + 12) + (2b^2c - 3bc - 8)$

 $15b^2c^2 - 2b^2c - 4bc + 4$

8. $(7e^2 + 3e + 2) + (9 - 6e + 4e^2) + (9e + 2 - 6e^2) + (4e^2 - 7e + 8)$

 $9e^2 - e + 21$

9. $(f^4g - fg^3 + 2fg - 4) + (3fg^3 + 3) + (4f^4g - 5fg) + (3 - 12fg^3 + f^4g)$

 $6f^4g - 10fg^3 - 3fg + 2$

10. Seis bloques de $4h + 4$ de altura cada uno y 3 bloques de $8 - 2h$ de altura cada uno están apilados uno sobre el otro para formar una gran torre. Halla una expresión para la altura total de la torre.

 $18h + 48$

Copyright © by Holt, Rinehart and Winston. All rights reserved.

LESSON 14-4

Practice

Subtracting Polynomials

Find the opposite of each polynomial.

1. $18xy^3$

 $-18xy^3$

2. $-9a + 4$

 $9a - 4$

3. $6d^2 - 2d - 8$

 $-6d^2 + 2d + 8$

Subtract.

4. $(4n^3 - 4n + 4n^2) - (6n + 3n^2 - 8)$

 $4n^3 + n^2 - 10n + 8$

5. $(-2h^4 + 3h - 4) - (2h - 3h^4 + 2)$

 $h^4 + h - 6$

6. $(6m + 2m^2 - 7) - (-6m^2 - m - 7)$

 $8m^2 + 7m$

7. $(17x^2 - x + 3) - (14x^2 + 3x + 5)$

 $3x^2 - 4x - 2$

8. $w + 7 - (3w^4 + 5w^3 - 7w^2 + 2w - 10)$

 $-3w^4 - 5w^3 + 7w^2 - w + 17$

9. $(9r^3s - 3rs + 4rs^3 + 5r^2s^2) - (2rs^2 - 2r^2s^2 + 6rs + 7r^3s - 9)$

 $2r^3s + 7r^2s^2 + 4rs^3 - 2rs^2 - 9rs + 9$

10. $(3qr^2 - 2 + 14q^2r^2 - 9qr) - (-10qr + 11 - 5qr^2 + 6q^2r^2)$

 $8q^2r^2 + 8qr^2 + qr - 13$

11. The volume of a rectangular prism, in cubic meters, is given by the expression $x^3 + 7x^2 + 14x + 8$. The volume of a smaller rectangular prism is given by the expression $x^3 + 5x^2 + 6x$. How much greater is the volume of the larger rectangular prism?

 $2x^2 + 8x + 8$ cubic meters

12. Sarah has a table with an area, in square inches, given by the expression of $y^2 + 30y + 200$. She has a tablecloth with an area, in square inches, given by the expression of $y^2 + 18y + 80$. She wants the tablecloth to cover the top of the table. What expression represents the number of square inches of additional fabric she needs to cover the top of the table?

 $12y + 120$ more square inches of fabric

Copyright © by Holt, Rinehart and Winston. All rights reserved.

LECCIÓN 14-4

Práctica

Cómo restar polinomios

Halla el opuesto de cada polinomio.

1. $18xy^3$

 $-18xy^3$

2. $-9a + 4$

 $9a - 4$

3. $6d^2 - 2d - 8$

 $-6d^2 + 2d + 8$

Resta.

4. $(4n^3 - 4n + 4n^2) - (6n + 3n^2 - 8)$

 $4n^3 + n^2 - 10n + 8$

5. $(-2h^4 + 3h - 4) - (2h - 3h^4 + 2)$

 $h^4 + h - 6$

6. $(6m + 2m^2 - 7) - (-6m^2 - m - 7)$

 $8m^2 + 7m$

7. $(17x^2 - x + 3) - (14x^2 + 3x + 5)$

 $3x^2 - 4x - 2$

8. $w + 7 - (3w^4 + 5w^3 - 7w^2 + 2w - 10)$

 $-3w^4 - 5w^3 + 7w^2 - w + 17$

9. $(9r^3s - 3rs + 4rs^3 + 5r^2s^2) - (2rs^2 - 2r^2s^2 + 6rs + 7r^3s - 9)$

 $2r^3s + 7r^2s^2 + 4rs^3 - 2rs^2 - 9rs + 9$

10. $(3qr^2 - 2 + 14q^2r^2 - 9qr) - (-10qr + 11 - 5qr^2 + 6q^2r^2)$

 $8q^2r^2 + 8qr^2 + qr - 13$

11. El volumen de un prisma rectangular, en metros cúbicos, está dado por la expresión $x^3 + 7x^2 + 14x + 8$. El volumen de un prisma rectangular más pequeño está dado por la expresión $x^3 + 5x^2 + 6x$. ¿Cuánto mayor es el volumen del prisma rectangular más grande?

 $2x^2 + 8x + 8$ metros cúbicos

12. Sara tiene una mesa cuya área, en pulgadas cuadradas, está dada por la expresión $y^2 + 30y + 200$. Tiene un mantel con un área, en pulgadas cuadradas, dada por la expresión $y^2 + 18y + 80$. Sara quiere que el mantel cubra la mesa. ¿Qué expresión representa la cantidad de pulgadas cuadradas de tela adicional que necesita para cubrir la mesa?

 $12y + 120$ pulgadas cuadradas adicionales de tela

Copyright © by Holt, Rinehart and Winston. All rights reserved.

Copyright © by Holt, Rinehart and Winston. All rights reserved.

LESSON **14-5**

Practice
Multiplying Polynomials by Monomials

Multiply.

1. $(x^2)(-3x^2y^3)$ — $-3x^4y^3$

2. $(-9pr^4)(p^2r^2)$ — $-9p^3r^6$

3. $(2st^9)(-st^2)$ — $-2s^2t^{11}$

4. $(3efg^2)(-3e^2f^2g)$ — $-9e^3f^3g^3$

5. $2q(4q^2 - 2)$ — $8q^3 - 4q$

6. $-x(x^2 + 2)$ — $-x^3 - 2x$

7. $5m(-3m^2 + 2m)$ — $-15m^3 + 10m^2$

8. $6x(-x^5 + 2x^3 + x)$ — $-6x^6 + 12x^4 + 6x^2$

9. $-4st(st - 12t - 2s)$ — $-4s^2t^2 + 48st^2 + 8s^2t$

10. $-9ab(a^2 + 2ab - b^2)$ — $-9a^3b - 18a^2b^2 + 9ab^3$

11. $-7v^2w^2(vw^2 + 2vw + 1)$ — $-7v^3w^4 - 14v^3w^3 - 7v^2w^2$

12. $8p^4(p^2 - 8p + 17)$ — $8p^6 - 64p^5 + 136p^4$

13. $4x(-x^2 - 2xy + 3)$ — $-4x^3 - 8x^2y + 12x$

14. $7x^2(3x^2y + 7x^2 - 2x)$ — $21x^4y + 49x^4 - 14x^3$

15. $-4t^3r^2(3t^2r - t^5r - 6t^2r^2)$ — $-12t^5r^3 + 4t^8r^3 + 24t^5r^4$

16. $h^2k(2hk^2 - hk + 7k)$ — $2h^3k^3 - h^3k^2 + 7h^2k^2$

17. A triangle has a base of $4x^2$ and a height of $6x + 3$. Write and simplify an expression for the area of the triangle.

$12x^3 + 6x^2$

LECCIÓN **14-5**

Práctica
Cómo multiplicar polinomios por monomios

Multiplica.

1. $(x^2)(-3x^2y^3)$ — $-3x^4y^3$

2. $(-9pr^4)(p^2r^2)$ — $-9p^3r^6$

3. $(2st^9)(-st^2)$ — $-2s^2t^{11}$

4. $(3efg^2)(-3e^2f^2g)$ — $-9e^3f^3g^3$

5. $2q(4q^2 - 2)$ — $8q^3 - 4q$

6. $-x(x^2 + 2)$ — $-x^3 - 2x$

7. $5m(-3m^2 + 2m)$ — $-15m^3 + 10m^2$

8. $6x(-x^5 + 2x^3 + x)$ — $-6x^6 + 12x^4 + 6x^2$

9. $-4st(st - 12t - 2s)$ — $-4s^2t^2 + 48st^2 + 8s^2t$

10. $-9ab(a^2 + 2ab - b^2)$ — $-9a^3b - 18a^2b^2 + 9ab^3$

11. $-7v^2w^2(vw^2 + 2vw + 1)$ — $-7v^3w^4 - 14v^3w^3 - 7v^2w^2$

12. $8p^4(p^2 - 8p + 17)$ — $8p^6 - 64p^5 + 136p^4$

13. $4x(-x^2 - 2xy + 3)$ — $-4x^3 - 8x^2y + 12x$

14. $7x^2(3x^2y + 7x^2 - 2x)$ — $21x^4y + 49x^4 - 14x^3$

15. $-4t^3r^2(3t^2r - t^5r - 6t^2r^2)$ — $-12t^5r^3 + 4t^8r^3 + 24t^5r^4$

16. $h^2k(2hk^2 - hk + 7k)$ — $2h^3k^3 - h^3k^2 + 7h^2k^2$

17. Un triángulo tiene una base de $4x^2$ y una altura de $6x + 3$. Escribe y simplifica una expresión para determinar el área del triángulo.

$12x^3 + 6x^2$

LESSON **14-6**

Practice
Multiplying Binomials

Multiply.

1. $(z + 1)(z + 2)$ — $z^2 + 3z + 2$

2. $(1 - y)(2 - y)$ — $2 - 3y + y^2$

3. $(2x + 1)(2x + 4)$ — $4x^2 + 10x + 4$

4. $(w + 1)(w - 3)$ — $w^2 - 2w - 3$

5. $(3v + 1)(v - 1)$ — $3v^2 - 2v - 1$

6. $(t + 2)(2t - 2)$ — $2t^2 + 2t - 4$

7. $(-3g + 4)(2g - 1)$ — $-6g^2 + 11g - 4$

8. $(3c + d)(c - 2d)$ — $3c^2 - 5cd - 2d^2$

9. $(2a + b)(a + 2b)$ — $2a^2 + 5ab + 2b^2$

10. A box is formed from a 1 in. by 18 in. piece of cardboard by cutting a square with side length *m* inches out of each corner and folding up the sides. Write and simplify an expression for the area of the base of the box.

$18 - 38m + 4m^2$ in.^{2}

11. A table is placed in a 14 ft × 18 ft room so that there is an equal amount of space of width *s* feet all the way around the table. Write and simplify an expression for the area of the table.

$252 - 64s + 4s^2$ ft^2

12. A circular swimming pool with a radius of 14 ft is surrounded by a deck with width *y* feet. Write and simplify an expression for the total area of the pool and the deck. Use $\frac{22}{7}$ for pi.

$616 + 88y + \frac{22}{7}y^2$ ft^2

Multiply.

13. $(r - 2)^2$ — $r^2 - 4r + 4$

14. $(2 + q)^2$ — $4 + 4q + q^2$

15. $(p + 4)(p - 4)$ — $p^2 - 16$

16. $(3n - 3)(3n + 3)$ — $9n^2 - 9$

17. $(a + b)(a - b)$ — $a^2 - b^2$

18. $(4e - f)^2$ — $16e^2 - 8ef + f^2$

19. $(2y + z)^2$ — $4y^2 + 4yz + z^2$

20. $(9p - 2)(-2 + 9p)$ — $81p^2 - 36p + 4$

21. $(m - 1)^2$ — $m^2 - 2m + 1$

LECCIÓN **14-6**

Práctica
Cómo multiplicar binomios

Multiplica.

1. $(z + 1)(z + 2)$ — $z^2 + 3z + 2$

2. $(1 - y)(2 - y)$ — $2 - 3y + y^2$

3. $(2x + 1)(2x + 4)$ — $4x^2 + 10x + 4$

4. $(w + 1)(w - 3)$ — $w^2 - 2w - 3$

5. $(3v + 1)(v - 1)$ — $3v^2 - 2v - 1$

6. $(t + 2)(2t - 2)$ — $2t^2 + 2t - 4$

7. $(-3g + 4)(2g - 1)$ — $-6g^2 + 11g - 4$

8. $(3c + d)(c - 2d)$ — $3c^2 - 5cd - 2d^2$

9. $(2a + b)(a + 2b)$ — $2a^2 + 5ab + 2b^2$

10. Una caja se construye con una hoja de cartón de 1 pulg por 18 pulg recortando de cada esquina un cuadrado que mide *m* pulgadas de lado y doblando los costados hacia arriba. Escribe y simplifica una expresión para determinar el área de la base de la caja.

$18 - 38m + 4m^2$ pulg^2

11. Una mesa está ubicada en una habitación de 14 pies por 18 pies de forma tal que queda un espacio de *s* pies de ancho alrededor de la mesa. Escribe y simplifica una expresión para determinar el área de la mesa.

$252 - 64s + 4s^2$ pies^2

12. Una alberca circular con un radio de 14 pies está rodeada por una plataforma de *y* pies de ancho. Escribe y simplifica una expresión para determinar el área total de la piscina y la plataforma. Usa $\frac{22}{7}$ para π.

$616 + 88y + \frac{22}{7}y^2$ pies^2

Multiplica.

13. $(r - 2)^2$ — $r^2 - 4r + 4$

14. $(2 + q)^2$ — $4 + 4q + q^2$

15. $(p + 4)(p - 4)$ — $p^2 - 16$

16. $(3n - 3)(3n + 3)$ — $9n^2 - 9$

17. $(a + b)(a - b)$ — $a^2 - b^2$

18. $(4e - f)^2$ — $16e^2 - 8ef + f^2$

19. $(2y + z)^2$ — $4y^2 + 4yz + z^2$

20. $(9p - 2)(-2 + 9p)$ — $81p^2 - 36p + 4$

21. $(m - 1)^2$ — $m^2 - 2m + 1$

Copyright © by Holt, Rinehart and Winston.
All rights reserved.